ENGINEERING
TEAM MANAGEMENT

ENGINEERING TEAM MANAGEMENT

DAVID I. CLELAND, Ph.D
HAROLD KERZNER, Ph.D.

VNR VAN NOSTRAND REINHOLD COMPANY
New York

Library of Congress Catalog Card Number: 85-29455
ISBN 0-442-21803-6

Manufactured in the United States of America

Published by Van Nostrand Reinhold Company Inc.
115 Fifth Avenue
New York, New York 10003

Van Nostrand Reinhold Company Limited
Molly Millars Lane
Wokingham, Berkshire RG11 2PY, England

Van Nostrand Reinhold
480 Latrobe Street
Melbourne, Victoria 3000, Australia

Macmillan of Canada
Division of Gage Publishing Limited
164 Commander Boulevard
Agincourt, Ontario MIS 3C7, Canada

15 14 13 12 11 10 9 8 7 6 5 4 3 2

Library of Congress Cataloging-in-Publication Data

Cleland, David I.
Engineering team management.

Bibliography: p.
Includes index.
1. Engineering—Management. I. Kerzner, Harold.
II. Title.
TA190.C59 1986 620'.0068 85-29455
ISBN 0-442-21803-6

PREFACE

Contemporary engineering organizations are modifying their organizational and management systems design to accommodate the complexity, interdependence, and change characteristic of modern organizations. A key change that emerges is the use of organizational forms where a number of persons are assigned to an interorganizational team for accomplishing the work of the organization. We believe that this represents a fundamental change in the way we will manage organizations in the future. For example, at the proposed General Motors Saturn Plant a committee of 35 company officials and 64 union members developed new concepts for labor-management cooperative efforts. At the proposed Saturn Plant teams of 6 to 15 workers will be responsible for meeting production schedules and budgets, controlling absenteeism, and handling health and safety. Job assignments will be handled by the teams, thus eliminating the traditional seniority system. Workers will do short-term planning and allocate resources. Long-term planning will be accomplished by a Strategic Advisory Committee composed of union and company representatives.

Teams are used in contemporary organizations often as a temporary organizational expedient, to deal with ad hoc organizational problems and opportunities. Project management and project engineering teams have been used for decades in contemporary organizations. In the last several years other team forms have gained popularity such as:

- *Production teams,* where workers do their own work planning and control
- *Worker-management teams* (sometimes called worker involvement teams), where the two sides who had not worked together before are now working together in a synergistic and cooperative fashion in many areas of the enterprise
- *Product-design teams,* composed of manufacturing, reliability, design, and quality engineers augmented with purchasing, supplier, and service experts to facilitate the design of a product
- *Task forces* that deal with issues and opportunities that cannot be easily handled by the regular organization
- *New business development teams,* which are being used to develop new business opportunities

The use of teams to complement the more traditional organizational approaches is on the rise. Contemporary and future engineering managers will be called on to provide environments where "team management" will contribute to organizational productivity. If organizational teams are properly planned for and implemented, the results will be beneficial for the organization and for the people.

Engineers, scientists, and technology-based technicians play a key role in contemporary organizational teams. With the emergence of large, complex activities, the role of the engineer and the engineering manager is taking on a new look in the 1980s, and this new look can be expected to continue through the 1990s. In the past, engineers were promoted to engineering management positions based upon technical rather than managerial expertise, with the argument that technical personnel would rather work for a manager with strong technical skills than for one with strong managerial skills.

Today, companies are taking a hard look at past mistakes, such as promoting a good engineer to line management simply to offer him more money. Today's engineering activities are constrained not only by *technical performance,* but also by *budgets* and *schedules.* Effective management of these three constraints (as well as other constraints of customer relations, political, social, and legal barriers) requires emphasis on team management and interpersonal skills. Therefore, engineering managers of the future must demonstrate managerial proficiency in the management of engineering organizations and the management of engineering/scientific-based organizational teams.

The complexities of modern technical projects require that work be integrated effectively by a team. These integration functions require that work be accomplished through others in a team setting advocating the need for strong managerial knowledge, skills, and attitudes. The necessity for better management requires that engineers also be familiar with strategies for sound leadership, motivation, decision making, and team building.

To fulfill this need to educate engineers in team management knowledge, skills, and attitudes, *Engineering Team Management* is broken down into the following chapters:

Chapter	*Title*
1	The Ambience of Team Management
2	Communications
3	Leadership
4	Motivation
5	Planning and Organizing

6	Conflicts
7	Decision Making
8	Time Management, Stress, and Burnout
9	A Framework for Developing High-Performing Technical Teams

We hope this book will prove to be a pragmatic source of management guidance for engineering managers and professionals. In modern engineering and scientific organizations there is a need for both managers and professionals to become more proficient in both leadership and followership of technical activities. We believe that our efforts in writing this book will enable managers and professionals to accomplish that proficiency.

We are indebted to Claire Zubritzky and Patricia Ray for their assistance in preparing the manuscript. As always we appreciate the administrators of our respective schools who continue to provide us with the cultural ambience to pursue the generation and transmission of engineering management knowledge, skills, and attitudes.

David I. Cleland, Ph.D.
Harold Kerzner, Ph.D.

CONTENTS

Preface/v
1. The Ambience of Team Management/1
2. Communications/37
3. Leadership/77
4. Motivation/112
5. Planning and Organizing/162
6. Conflicts/196
7. Decision Making/227
8. Time Management, Stress and Burnout/264
9. A Framework for Developing High–Performing Technical Teams/297

Index/327

ENGINEERING TEAM MANAGEMENT

1
THE AMBIENCE OF TEAM MANAGEMENT[1]

Management is the discipline that deals with the judicious means to accomplish an end purpose. A *manager* is one who provides leadership for a group of people to cooperate in accomplishing an end purpose. The opportunity for management exists wherever people are trying to cooperate in accomplishing an end purpose. More specifically, management is an intellectual process that deals with

- Providing leadership for and working with a team of people
- Working toward an end purpose
- Establishing a suitable balance between human and nonhuman resources
- Making and implementing decisions
- Facilitating an environment in which people are willing to work together

Management is an ancient process acquired by experience, study, and observation. As a branch of learning, management involves the conscious use of acquired knowledge and creative imagination to provide leadership for people having a common purpose for working, living, and playing together. A manager is one who has the faculty of planning and carrying out expertly what is planned for an organization.

An organization is an entity of human and nonhuman resources working toward a common purpose. The management of organizations pervades every aspect of modern society. In contemporary organizations the need for management is found in those situations where some form of "teams" are used to complement an existing organizational design. A key

[1] Portions of this chapter have been paraphrased from David I. Cleland, "Matrix Management (Part II): A Kaleidoscope of Organizational Systems," *Management Review*, December 1981, pp. 48–56; and from David I. Cleland, "Matrix Management in the Engineering and Manufacturing Communities," paper presented at the International Congress on Technology and Technology Exchange, October 8–10, 1984, Pittsburgh.

Table 1-1. The Two Types of Engineering Teams.

TRADITIONAL	NONTRADITIONAL
Example: Engineering manager and staff	Example: Product design team or engineering project team
Team is permanent, formed by the pyramid organizational structure	Team is temporary, appointed for a specific purpose
Formal authority, responsibility, and accountability	Formal authority, responsibility and accountability may be defined; informal authority is important to the operation of the team
Single reporting relationship to a common supervisor	Dual reporting relationship, to an administrative supervisor and a team leader

overriding feature of this design is a departure from the traditional form of management in favor of a team form where there are multiple authority, responsibility, and accountability relationships resulting in a sharing of decisions, results, and rewards. Other features include the following:

- A team culture predominates as the organizational design through which objectives and goals are accomplished. These teams may be appointed by management or may arise through the interactions of the peer groups.
- Participative management is a way of life in the organization.
- Key ad hoc organizational problems/opportunities are dealt with through some form of temporary organizational design.

Pragmatic results from the use of teams in modern organizations are going hand-in-hand with a growing skepticism that a suitable organizational structure is the principal means to implement organization strategy. More and more companies are employing a complementary pyramidal and team structure, shifting attention to the human dimension of management.[2] By emphasizing the human dimensions through leadership and followership, complex and interdependent organizational problems and opportunities can be better addressed. As organizations grow and as competitive and environmental forces become more complex, structure becomes less effective as a mechanism for unifying organizational purposes

[2] Peters and Austin found that small scale via team organization and decentralized units are vital components of top performance. See Tom Peters and Nancy Austin, "A Passion for Excellence," *Fortune,* May 13, 1985, pp. 20–32.

or bringing about effective teamwork. Such intangibles as style, communication, shared values, and individual and team role performance count the most toward ensuring operational effectiveness. If these trends continue, future organizations viewed in structural terms will appear to be simpler but the management processes to effectively integrate many different team modes of operation will be complex and demanding.

Engineering teams are of two general types: *traditional* and *nontraditional.* Table 1-1 describes these types. In the sections that follow, an overview will be given of some of these alternative organizational designs found in contemporary organizations, starting with the best-known design: project management.

PROJECT MANAGEMENT

Project management, although practiced in some manner for centuries, emerged as a subdiscipline in the management field in an unobtrusive manner in the early 1960s. The terms *project* and *program* were used interchangeably to describe a management approach which already existed in an informal form in many organizations. No one can claim to have invented project management; its beginnings are often cited in the ballistic missile program of the United States. The origins of project management can be found in the management of large-scale construction projects, in the Manhattan Project, or in the use of naval task forces. Some writers have attempted to trace the origins of project management from antiquity, citing the building of the pyramids as one example of the management of a large ad hoc project. No one doubts the antiquity of the practice of project management. In modern times it is clearly established as an important part of contemporary management theory and practice.

Project-driven "matrix" organizational designs emerged in companies to satisfy several operating and strategic needs: resource sharing, profit center integration for large projects, customer requirements, competitive pressures, and the serving of specific market segments. Although the concept of a "matrix" organizational model emerged in practice, it was not until 1964 that the term was conceptualized in the literature by Professor John F. Mee at Indiana University. In describing the two-dimensional organizational model found when project teams are superimposed on an existing functional structure, Professor Mee noted:

> A matrix type of organization is built around specific projects. A manager is given the authority, responsibility, and accountability for the completion of the project in accordance with the time, cost, quality, and quantity provisions in the project contract. The line organization

develops from the project and leaves the previous line functions in a support relationship to the project line organization.[3]

Professor Mee's definition of the matrix organization set the tone for a proliferation of literature directed to describing the specificity of this organizational model. Project management provided a powerful leading edge to a family of alternative applications of matrix management found in use today. These alternative applications are discussed in the following section, starting with project engineering, the precursor of project management.

PROJECT ENGINEERING

Within the engineering community, project management first evolved out of the specialized activity of project engineering. Project engineers are usually responsible for directing and integrating all technical aspects of the design/development process. They manage a product through all of its engineering steps, from initial design to service life.

In order to accomplish this successfully, they must work closely with the other engineers who support the project. When a problem with the design/development of a product occurs in a functional department, the project engineer is responsible for working with that department to correct the problem. In such situations, questions of authority and responsibility are raised because the project engineer has become dependent on other people over whom he has limited control.

In some companies, the project engineering function is, in effect, the project management arm of the engineering department. In another context, project engineering is the business of building plants from the preliminary study through the design, procurement, erection, and trial operation.

Figure 1-1 portrays the responsibility of project engineers in directing the design/development activities within an aerospace company. Figure 1-1, which illustrates the project management context of project engineering, shows how the engineering disciplines and the engineering project teams interface in the matrix organization. The engineering personnel assigned to the design teams come from the different disciplines supporting the project. The "interface" of the vertical structure and the design team provide a focal point for pulling together the specialized engineering effort to support a project. The management of the engineering project is handled through designating an individual as a *project engineer* to provide

[3] John F. Mee, "Matrix Organization," *Business Horizons*, Summer 1964.

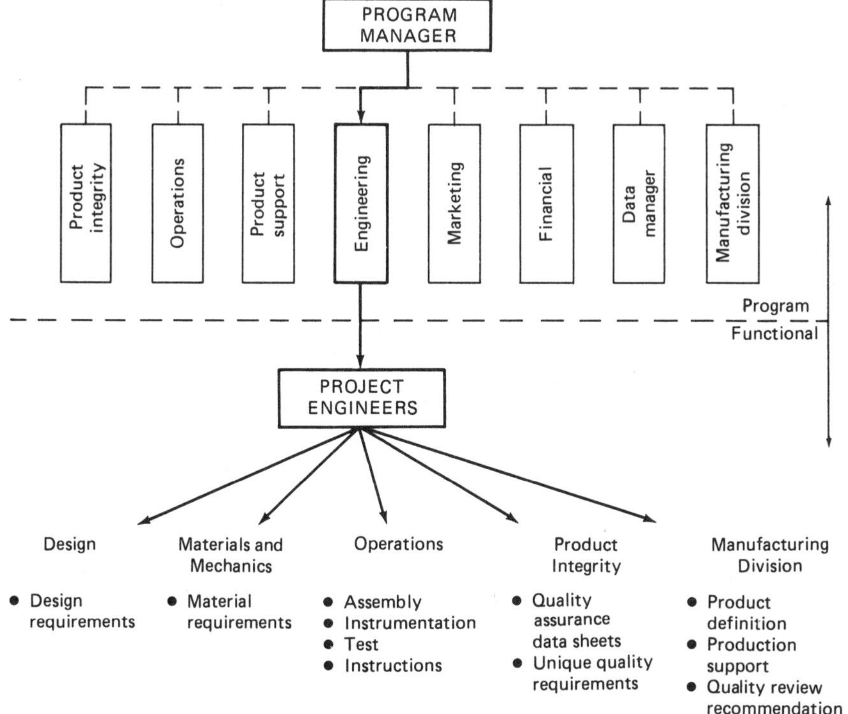

Figure 1-1. Project engineering in the program management context. Project engineers are responsible for directing the design development activities. (Adapted from *Design/Development Process Guide,* Pratt & Whitney Aircraft Group, Government Products Division Subsystem Guide 3-3, United Technologies Corporation, 1978.)

an accountable focal point for pulling all needed specialized interests into focus on the project. The focus of the specialized support on the project comes together in terms of an *authority-responsibility matrix.*

The organizational model reflected in Figure 1-2 is used to represent this authority-responsibility matrix in plant building. In this figure the specialized functional groups—the engineering disciplines—interface with the projects as organizational resources are applied to these projects.

Figures 1-1 and 1-2 reveal the kinship between two kinds of project engineering and project management. Both fit the description of matrix management conceptualized by Mee.

Project management and project engineering are the most familiar forms of matrix management. However, other forms are coming into use that fall under the general category of worker-management cooperative teams, such as production teams.

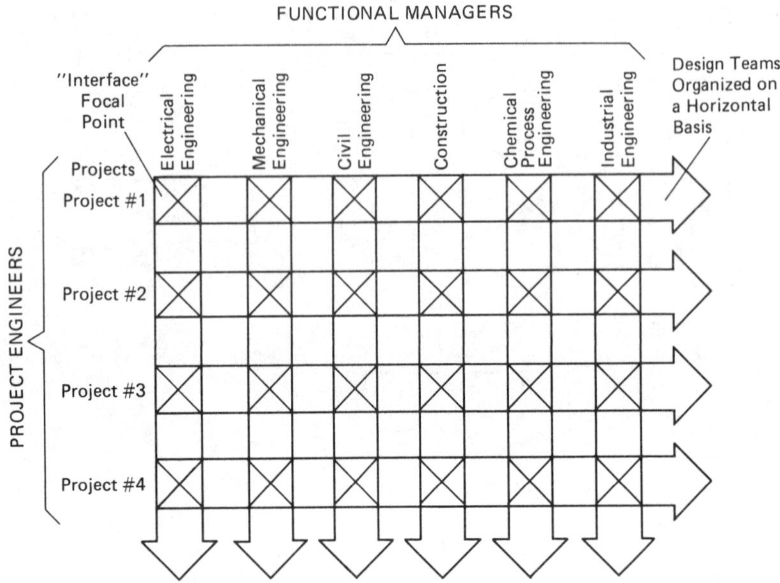

Figure 1-2. Matrix organization for engineering design of an industrial plant. (*Source:* David I. Cleland and Dundar F. Kocaoglu, *Engineering Management* (New York: McGraw-Hill, 1981, p. 34.)

PRODUCTION TEAMS

In manufacturing, production teams of workers do their own work planning and control. In such a setting, the supervisor becomes a facilitator who helps the teams work out the details of assuming responsibility for the manufacture of the product. In different plants, however, these teams operate in different ways.

In some companies the pattern for production teams was set decades ago. For example, at the Lincoln Electric Company in Euclid, Ohio, worker representation in the management of the company was started in 1914. Workers elect representatives from each department on an advisory board that meets twice a month to advice senior executives. Suggestions by this group have helped shape the company's success.[4] The use of workers to advise management at Lincoln is just one of many unique worker-management relationships in this company. Others include the following:

[4] William Serrin, "The Way That Works at Lincoln," *New York Times*, January 15, 1984, p. 4F.

- Guaranteed employment for all full-time workers with more than two years' service
- No mandatory retirement
- High wages, including an annual bonus based on the company's profits
- Pay based on what is produced rather than hourly or weekly wage
- Promotion almost exclusively from within
- A supervisor-to-worker ratio of 1 to 100
- Mandatory overtime and job assignments
- No unions
- Employee stock purchase plan; about 75% of workers own 40% of the stock
- Turnover of less than 3% a year to include retirements

No other sizable company has so many worker-management features as Lincoln Electric. Although the company insists on individual initiative and pays according to individual effort, the notion of team synergy and cooperative group work is typical of what has been described as "The Lincoln Family."[5]

A strong impetus for more employee participation through service on teams came in 1980, when the United Steelworkers and several large steelmakers that bargain as a group agreed to form "labor-management participation" teams in selected plants. This initial experiment has moved past the design and development stage and has resulted in better labor relations, improved quality and efficiency, and reduced absenteeism and grievance loads. Acceptance of team participation has not been unanimous; some workers fear that it is a ploy to manipulate them. First-level supervisors fear that the teams will undermine their authority. Members of the participation teams feel pressure from peer groups, both from team members and from nonjoiners, often of a critical nature. Prudent management of these teams recognizes that such peer pressure is to be expected, and if properly managed, can provide valuable "checks and balances."

TRW Systems Corporation created semiautonomous work teams in one of its manufacturing plants. The workers assemble a product on a team basis rather than performing assembly line tasks separately. Teams are allowed to schedule their own time as long as they do the job.

The Volvo auto company has one auto assembly plant at Kalmar, Sweden, where workers are organized into groups of six or more persons working as teams. For example, one team or workers injects a sealing compound in all seams and installs sound-dampening insulation. They

[5] Ibid.

rotate the team's 15 functions so that no one gets fatigued or bored performing the same job. As Pehr Gustaf Gyllenhammer, the chief executive to Volvo contends, "We have to change the organization so the job itself provides more for the individual. We will never build another production line as long as I am in command at Volvo."

At GM's Fisher Body Plant No. 2 in Grand Rapids, Michigan, the 2000 employees have been organized into six business teams, each team being essentially a business unto itself with its own maintenance, scheduling, and engineering personnel. Currently, not only salaried employees at all levels participate in deciding how to meet their team's objectives, but hourly-wage team members are included in the decision-making process as well.

Sometimes the formation of a worker-management team arises out of a crisis. At the Detroit Trim plant of the Chrysler Corporation, a planned closing stimulated the union and Chrysler management to work together to keep the plant open. The two sides—who had not worked together before in a synergistic and cooperative fashion—fashioned an agreement that completely altered the economics of the plant, the way it works, and the way it is managed. Some of the more important changes that were implemented included

- Reduction of work force by 25%
- New production quotas
- Flexibility in job assignments
- Smaller work teams
- Broadened managerial responsibilities
- Permanent labor-management cost control panel

This plant achieved extraordinary increases in output, primarily by getting a full day's work out of its employees. A willingness to work together as a team, to share information, and to build mutual respect and trust is what kept this plant open.[6]

Some plants use production teams to raise the quality of the product. At the Honda Motorcycle plant near Columbus, Ohio, quality is higher than that of Japan's comparable product.[7] In the early months of the plant's operation the company spent a lot of time training its employees in the kind of teamwork that achieves high quality. Work attendance ranks favorably with that of the workers in Honda's plants in Japan. Comparative productivity figures are not disclosed, although the vice-president of

---

[6] Jeremy Main, "Anatomy of an Auto-Plant Rescue," *Fortune*, April 4, 1983, pp. 108–113.
[7] Masayoshi Kanabayashi, "Honda's Accord," *Wall Street Journal*, October 2, 1981, p. 1.

Honda of America says that in Honda's experience, ". . . there is no difference at all" between Japanese and U.S. workers.[8]

Part of the success at this plant may come from the "people-oriented" management approach of the Japanese. The company executives go out of their way to play down status distinctions between workers and managers and to treat workers as members of a team. Everyone eats in the same cafeteria, no one has reserved parking places, and employees are called "associates."[9]

At the Cadillac engine plant in Livonia, Michigan, workers and supervisors cooperate closely on "business teams," composed of 10 to 20 workers, that plan and organize the work and make decisions previously left to management. Unlike the traditional assembly line where a few simple tasks are performed, here each worker learns to handle several jobs. Since all the members of a team learn all the jobs in one section, more flexibility is achieved for work assignments and for replacement of absent workers. Employees work and "supervise" each other in small teams at "work islands" where jobs are rotated and socializing is permitted. Team members meet weekly to deal with such matters as safety, housekeeping, pay raises, job rotation, or even the redesigning of the work flow.[10]

In some organizations the work team is taking on nontraditional duties. For example, at Romac Industries, a small nonunion rubber-products manufacturer in Seattle, workers are allowed to propose wage hikes for themselves, subject to the approval by vote of their peers. Managers keep track of the plant's efficiency; if a week's production falls below the average for the previous four weeks, there is a moratorium on raises. If it exceeds that standard, then the workers can choose between paid time off and a period of work at double pay.[11]

In the steel industry, where labor relations with the United Steelworkers (USW) have been traditionally hostile and managers have practiced an authoritarian management style, labor-management participation teams are now being used. These participation teams permit workers to help solve production problems. At the Jones & Laughlin Steel Company, one team developed a technique for removing surface defects from steel slabs, saving the company $1.0 million a year.[12]

At the Ford Motor Company under an "Employment Involvement Program" (EI) the United Auto Workers (UAW) and the company are striv-

---

[8] Ibid.

[9] Ibid.

[10] "A Work Revolution is U.S. Industry," *Business Week,* May 16, 1983, p. 108.

[11] "Special Report, The U.S. Productivity Crisis," *Newsweek,* September 8, 1980, pp. 50–69.

[12] "Time Runs Out for Steel," *Business Week,* June 24, 1983, pp. 90–91.

ing to make work more satisfying and productive. Over 8000 workers are involved in various problem-solving committees on the plant floor. There are EI programs at the majority of Ford plants and offices; union-management steering committees are in place at all 65 assembly plants.[13]

At Crane Plastics Inc., in Columbus, Ohio, matrix management was introduced as the company struggled through a difficult transition in its manufacturing operation. A vice-president of a "matrix" profit center was appointed to manage 7 to 10 teams that are usually in operation. As part of its strategy to convert to large-volume custom plastic products, matrix management techniques are used to pull together several different disciplines to produce a customized product. Before matrix techniques were introduced, the modus operandi is described as follows:

> People would have a meeting, . . . and they'd have their assignments. They'd come back after two or three weeks with good intentions, but they just didn't get it done, because they had more important things to do in their own functional areas and, . . . there was no one to say, "you didn't do what you were supposed to." *They were peers without a leader.*[14]

Toyota and General Motors (GM) have a joint venture to build subcompact cars at the old GM assembly plant in Fremont, California. Executives of the new joint-venture company, New United Motor Manufacturing Inc., have been designing a bold experiment in team management: a labor relations system so dependent on worker-manager cooperation that "the very line between union and management has blurred."[15] A key provision stipulates that all assembly workers are to be covered by a single job classification and organized into five- to seven-man work teams. Each team member will be trained to work at all jobs done by the team, and eventually will be trained to work at other team jobs in the same general area. Team members will be expected to solve day-to-day problems. The salaried foremen perform as "facilitators." They will act as mediators, working with five to seven teams. By working with managers of the functional line departments such as welding, painting, and assembly, they will provide and coordinate resource support to the teams. The plant will have only five layers of management, fewer than the typical U.S. plant.

---

[13] Reported in John Hoerr, "Smudging the Line Between Boss and Worker," *Business Week,* March 1, 1982, p. 91. See also, "What's Creating an 'Industrial Miracle' at Ford," *Business Week,* July 30, 1984, pp. 80–81.

[14] Ellen Kolton, "Team Players," *INC.*, September 1984, p. 140. Emphasis added.

[15] Michael Brody, "Toyota Meets U.S. Auto Workers," *Fortune,* July 9, 1983, pp. 54–64.

The challenge at this plant is to get the workers, the union, and the management wholeheartedly committed to working together more efficiently in the performance of both individual and team operations. The use of work teams, accompanied by a commitment from all people for success, should help to meet the challenge.

Recently, Deere and Company, a large farm equipment manufacturing plant, in cooperation with the United Auto Workers, launched its first formal employee participation program to facilitate worker-management team effort. The objective of this program was to develop more participation by employees in the factory's operation, giving them an opportunity for direct input into the management process as it influences the individual's work area. The expectation for this program is that it will ultimately lead to improved competitiveness through greater job satisfaction, increased productivity, and improved product quality. The importance this company places on employee participation is keynoted in the chairman's comments in the Annual Report:

> We all know that people make the difference. If we can provide them with a better climate to use their minds, then I'm confident that U.S. industry will not only improve, but it will become more competitive in sectors now under pressure and enhance its leadership in areas where we're already strong. At the same time we believe all employees will derive much greater satisfaction from the work they do.[16]

With such a positive attitude toward employee participation, the chances for unique work teams to be successful at Deere & Company seem high.

Production teams are sometimes described as "semiautonomous work groups." Semiautonomous work groups are given responsibility for performing a variety of tasks, for making decisions that affect their jobs, and for resolving day-to-day authority problems that may arise.[17] The number of people in a group varies from a few to as many as 20 individuals. The operating procedures include such things as

- Interchanging tasks among themselves
- Setting, adjusting, maintaining, and repairing their equipment
- Requisitioning and inspecting incoming material
- Inspecting finished products

---

[16] Robert A. Hanson, *Deere & Company 1983 Annual Report*, John Deere Road, Moline, Illinois 61265, p. 19.

[17] Panagiotis N. Fotilas, "Semi-Autonomous Work Groups: An Alternative in Organizing Production Work?" *Management Review*, July 1981, pp. 50–54.

- Reporting work progress
- Maintaining safety
- Training new members
- Administering discipline

What degree of autonomy to give such groups is a key management decision. One authority in the operation of such groups notes the following:

> One of the most delicate and challenging tasks facing management in setting up work groups is reaching the decision on the degree of autonomy and the range of responsibility and authority to be given to such groups.[18]

The introduction of such groups changes the production and organizational systems with which managers and first-level supervisors are familiar. Time is essential in preparing everyone for the change. People must understand how these groups will affect such matters as production costs, delegation of authority, seniority, layoff policy, decision making, and acceptance of responsibility, and how important it is that the people are committed to making the groups work effectively.

At the Packard Electric Division of General Motors, the worker participation process, which involves thousands of workers in problem-solving groups and self-managed work teams, has dramatically improved the union-management relationship. Local union officials are involved in manpower decisions such as hiring, job assignments, transfers, and layoffs; union officials are indeed a part of management.[19]

ORGANIZING WORKER-MANAGEMENT TEAMS

Worker-management teams may be formed and organized in different ways:

- A company initiates changes in its policies and gradually involves the union in further changes.
- A joint labor-management committee forms and agrees to a philosophical statement of cooperation.
- In rare instances the union launches the initiative and convinces management to join up.

---

[18] Ibid., p. 53.

[19] "A Pioneering Pact Promises Jobs for Life," *Business Week*, December 31, 1984, pp. 48–49.

- Management and unions go together to a retreat and discuss what such cooperation implies for their situation.
- A union and management start by working together on less sensitive topics that are unlikely to spawn conflict.[20]

The chief factor in the formation of these work teams is the realization by union and management that cooperation has a value of its own, unrelated to other aspects of the team relationship.

One of the most challenging parts of the success of these teams is the need to bring about a major change in the role of first-level supervisors. The supervisor becomes a "facilitator" for arranging for resources for the work team, expediting the many participative processes needed to maintain the cohesiveness and effectiveness of the team, counseling and instructing workers, helping to establish group objectives, and acting as an integrator-communicator for the team. Therefore, inputs keynoting positive attitudes to cooperate must come from the supervisors as well as the employees.

Success with worker-management teams depends on how the purposes or results of the work team are handled. The team members must have input on changes they must live with. Two fundamental steps must involve them: *First,* the design of appropriate technologies and processes for whatever the teams wish to accomplish; the second, the implementation of the technologies and processes. If these steps are not closely integrated with the full understanding and support of the members of the team, it will probably lead to union-management mistrust. Nothing kills the synergy of a team and the initiative of the members more than unilateral decisions for major changes in technology, processes, or strategy. It must be kept in mind, however, that labor-management teams rarely meet with early success. The team must exist under the ambience of an adverse relationship that has been characteristic of the past. It is only when both parties learn to develop an attitude of trust and dignity in trying to make the team work that adverse relationships can be reduced and success becomes more probable.

Doubt, according to Scobel, is the worst enemy of labor-management cooperation. But teamwork, once rooted, is difficult to uproot.[21] Sufficient experience with labor-management team effectiveness has shown the way to improved productivity, security, and growth, both individual and organizational. The "revolution" to more labor-management cooper-

[20] Donald N. Scobel, "Business and Labor—From Adversaries to Allies," *Harvard Business Review,* November–December 1982, pp. 129–136.
[21] Ibid., p. 136.

ation through teams will not collapse; it may retreat, but the changes that have resulted will not disappear.

PRODUCT-DESIGN TEAMS

In some manufacturing companies, product-design teams are used to formulate design concepts. The intent of these teams is to provide an opportunity for everyone with responsibility for a product to critique a design before it is finalized. Such teams include reliability engineers, manufacturing engineers, purchasing representatives, suppliers of purchased parts, and service experts. When a team approach is used, the various functions of manufacturing are taken into account at the very outset of the design process. In addition to facilitating the design of a product, these teams ensure that the product meets the desired quality standards. Since what is designed eventually will be manufactured and serviced, even a small change in design may result in a better product: one that is more reliable or easier to service.

But the involvement of manufacturing engineers in the design process has quality implications far beyond their suggestions for minor design changes. In the experience of one company, their inclusion has provided

> an early warning of potential manufacturing problems that may be involved with new designs so they can develop adequate safeguards. As an example, the manufacture of a single-piece, thin-walled drive shaft in the 15-speed power shift transmissions used on the company's new line of tractors was identified as a potential problem early in the design process. The concept was a radical departure from old designs and involved manufacturing techniques not normally associated with the production of shafts. Failure would involve high service costs. Working closely together, the product designers and manufacturing engineers developed ideas for 12 improvements in the manufacturing process of these shafts. The ideas were implemented prior to the start-up of manufacturing and the new shaft has been produced without problems.[22]

QUALITY CIRCLES

Quality circles (QCs) are composed of small groups of employees who participate in improving their work and work environment. The quality

[22] *Deere & Company 1983 Annual Report,* John Deere Road, Moline, Illinois 61265.

circle is a productivity and quality improvement technique that was implemented in Japan in the early 1960s. The formation of QCs in Japan grew out of the need to change the inferior image of the Japanese product. The Japanese introduced the concept of having the responsibility for quality reside with each member of the organization. Training in quality control was therefore extended both to the supervisory and to the worker level in the organization.

A quality circle is a group of people (4 to 10) with a common interest who meet regularly to participate in the solution of job-related problems and opportunities. Operating in the work environment, it is an ongoing group that performs "opportunistic surveillance" for the organization: searching for opportunities, defining problems by applying formal data collection and analysis, and arriving at solutions that are presented for acceptance and implementation by management.

In 1973 the quality control circle concept was implemented in the Lockheed Missile and Space Company. Today it is estimated that quality control circles exist in some 2000 American companies such as General Motors, Westinghouse Electric, Ford, RCA, Bowing, International Harvester, American Airlines, and Rockwell International.

TASK FORCE MANAGEMENT

Task forces are used by companies to deal with problems and opportunities that cannot be easily handled by the regular organization. Usually these problems or opportunities cut across organizational boundaries. A task force can be a powerful mechanism for bringing talent to focus on complex matters. When the objective for which the task force was organized is attained, the group disbands.

A task force composed of persons drawn from appropriate elements of an organization deals with a short-term problem or situation. People are usually assigned to work on a task force on a part-time basis and find that they have to satisfy two bosses during this period–their regular supervisor and the task force leader. Since each member usually represents a different part of the organization and brings different viewpoints, goals, loyalties, and attitudes to the group, the job of integrating individual efforts is no small challenge for the leader.

Task force experience is an important contribution to a professional's career. For example, in his book *Designing Complex Organizations* (Reading, Mass.: Addison-Wesley, 1973), Jay R. Galbraith suggested that to be effective, a task force should meet these criteria:

- Members must be knowledgeable representatives of their organizational needs.
- Members must be in a position to speak with authority on behalf of their organization.
- Members must have the skills necessary for making decisions on the basis of information and expertise and must not be preoccupied with questions of formal authority.
- Generally, task force activity should be a substantial job obligation for only a few members, while the majority remain full-time members of their home department with only a part-time, albeit meaningful, commitment to the task force."[23]

Sometimes task forces are used to help change an organization's cultural ambience. But without careful preparation, problems may occur. For example, Boyle describes the experiences of employees in one division that attempted to change the company's traditional militaristic style of management and substitute a more participative one.[24] The author recalls that this change was "fraught with unanticipated problems and issues." The division had launched a series of six official task forces to facilitate the change to participative management. Eventually some 200 other task forces were, in the words of the author, giving ". . . our first clue that the whole affair might be running amok." The six official task forces were not being managed properly. Lack of direction, unclear reporting relationships, and vague operating procedures were some of the issues that were observed in the operation of the task forces. In a candid retrospective view the author cites some of the following mistakes that were made:

- There were no ground rules for the task forces or for the management committee to whom the task forces reported.
- There was no plan of action for the task forces.
- The task forces were isolated from the main business of the company, with little coordination between the task forces.
- Some task forces became too self-serving, forgetting that they were to represent the whole organization.
- There was difficulty in relating the time spent on task force activity to its beneficial impact on the organization.

---

[23] Ibid.

[24] Richard J. Boyle, "Wrestling with Jellyfish," *Harvard Business Review,* January–February 1984, pp. 74–83.

The lesson to be learned from the experiences at this company: *Task forces have to be well managed to be effective.*

However, some companies have had more positive results. The participation teams used by the Honeywell Defense and Marine Systems Group eventually became a *method* for achieving participative results in an effective team, not simply an end in itself.[25]

Ware suggests that task forces are appropriate when organizations have situations or problems which have the following characteristics:

- Several different functional areas are or will be affected.
- The problem is relatively urgent.
- The existing organization is not equipped to address the problem.
- The nature of the solution is highly uncertain.
- The solution will potentially involve significant changes in organizational structure, responsibilities, or operating systems.[26]

Task forces are, according to Ware, most commonly established in response to some organizational crisis.

CRISIS MANAGEMENT TEAM

The matrix structure has been suggested as one to use in managing a crisis. In crisis management the purpose is to utilize a lean, cost-effective organizational design that represents the minimum burden to the organization and provides a flexible quick response team capability. Team members are drawn from the various functional organizations, providing the expertise required to deal with the crisis that is under way. The American Management Association has suggested the use of a team to manage a crisis. The crisis management model suggested by AMA is depicted in Figure 1-3.

Crises in business organizations can take many forms, such as:

- Fire and natural disasters
- Energy shortages
- Economic downturn

[25] Ibid.

[26] James P. Ware, "Making the Matrix Come Alive: Managing a Task Force," in David I. Cleland (ed.), *Matrix Management Systems Handbook* (New York: Van Nostrand Reinhold, 1983), p. 114.

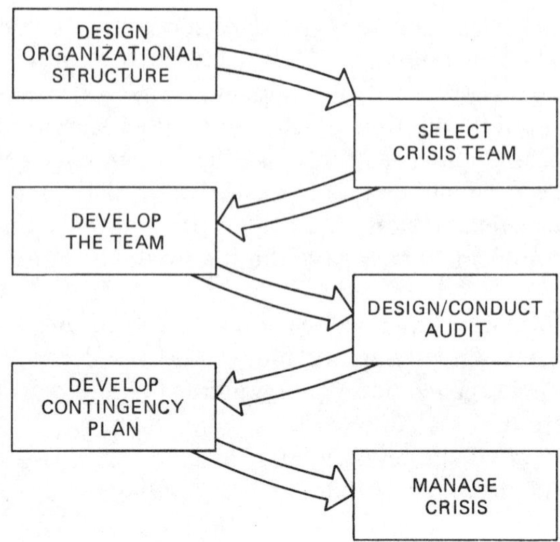

Figure 1-3. Crisis management mode. (Reprinted, by permission of the publisher, from *Crisis Management: a Team Approach*, by Robert F. Littlejohn, an AMA Management Briefing, p. 14 © 1983 AMA Membership Publications Division, American Management Associations, New York. All rights reserved.)

- Theft
- Product recalls
- Legal action
- Corporate takeover attempt

TEAM MANAGEMENT

A successful team management program requires some basic changes in the management philosophies of the managerial staff. First, the team must be recognized as an integral unit of the organization. Being so recognized, such a team requires the elements of management that an organization must have in to order succeed: a mission, objectives, goals, and strategy supported by an appropriate structure; role definition; leader and follower style; organizational system resource support. If these elements are provided, the team has the basics to be effective.

Second, the key managers must be responsive to the needs of the team members to truly participate in the decision processes in the organization. This requires that managers develop the interpersonal skills and foster a culture that is supportive of a legitimate and explicit role in the organizational decision process. If managers truly encourage employee participation by providing opportunities for problem solving and decision making, a team supportive culture has a much better change of developing.

Third, management must work at communicating, providing team members the information they require to be effective and providing continuing feedback on how well they are doing. Of course, the team members must also work at communicating so that both managers and team members have adequate information to make informed decisions.

Fourth, managers and team members jointly have to work at building trust through the words they use, the deeds they perform, the ideas they put forth, and the basic beliefs and philosophies they promote. In other words, they must build trust through a demonstrated basic belief and guiding philosophy practiced and articulated in their environment. In their beliefs and philosophies they accept that people are, to paraphase Douglas McGregor, basically OK, can be self-managed, will build confidence, are trustworthy, and enjoy meaningful work.

Since teams are a special form of organization, pretty much the same things that make an organization effective will make a team effective. An important part of effective team management depends on how well the strategy for the team's operation has been developed and implemented. There are several important factors that require definition and development to help facilitate the effectiveness of the team.

- A clear, concise statement of the *mission* or *purpose* of the team. This should be a statement of the "business" the team is in.
- A summary of the *goals* or *milestones* that the team is expected to accomplish in working toward its mission or purpose.
- A meaningful identification of the principal *tasks* required to accomplish the team's purposes.
- A summary delineation of the *strategy* of the team to include major policies, programs, procedures, plans, budgets, and other resource allocation methods required to facilitate its operation.
- A statement of the team's organizational design—basically a summary statement of the *roles* of authority, responsibility, and accountability that people are expected to fill.
- A description of the kind of leader and follower *style* required to further team integrity, cohesiveness, performance, and well-being.
- A identification of human and nonhuman *resource* support.

Leadership of the team is critical. A team leader is a manager in the truest sense of the work, carrying out the management functions of planning, organizing, motivating, directing, and controlling. In managing a team, however, the leader may be judged by his or her ability to get the members of the team to communicate with each other more than anything else. As a result, most of the leader's time is spent in some form of communication with the team's members to satisfy their needs for information on the team's work and how the members are to work together. Team building is often carried out to enhance the effectiveness of the team.

Team building is the process of creating and then maintaining effective team functioning. Because there is usually a strong link between the interpersonal relations among team members and the team's performance, team building emphasizes both.[27] The use of consensus techniques within a group can help to facilitate the team's operation.

Consensus has been defined by Hall as a condition in which each member accepts the team's decisions because they seem most logical and feasible. He suggests several guidelines to assist team members to achieve consensus:

- State your position as clearly and logically as you can—but do not argue for it. Listen to and ponder the other members' reactions before you push your point.
- If discussion between some members gets bogged down on any one point, do not treat it as a win-or-lose proposition. Instead, seek out the next most acceptable alternative.
- Do not yield on any point just for the sake of harmony. Accept a solution only when it is based on sound logic.
- Shun techniques that bypass logic for the sake of reducing conflict (such as majority vote, flipping a coin, bargaining, and averaging). When a dissenting member finally goes along with the group, don't make up for this by letting the yielder have his or her own way on some other point.
- Root out differences of opinion and pull everyone into the discussion. Only by airing the widest possible range of opinions and drawing in all available information can the group come up with high-quality solutions.[28]

Hall found that teams using these guidelines performed better in solving problems than teams not so trained. Through the use of team problem-

[27] David Wilemon, "Developing High-Performance Matrix Teams," in David I. Cleland (ed.), *Matrix Management Systems Handbook* (New York: Van Nostrand Reinhold, 1984), pp. 363–378.

[28] Jay Hall, "Decisions, Decisions, Decisions," *Psychology Today*, November 1971, p. 86, cited in James A. F. Stoner, *Management* (Englewood Cliffs, N.J.: Prentice-Hall, 1982), p. 346.

solving and consensus decision-making techniques, power equalization takes place, which in turn helps to build trust.

A professional's first experience in managership is usually in some team context. At this time the professional begins to wonder what management is all about.

THE PROFESSIONAL'S TRANSITION TO MANAGEMENT[29]

The professional's first experience as a manager will probably be one of two types:

1. As a leader of a team formed for some specific purpose in the organization
2. As a first-level supervisor of people providing specialized services to an organization such as programming, accounting, engineering design, and so forth

A supervisor is sometimes called a "working manager," meaning that he works along with specialists who are his "subordinates" and for whom he is responsible. The general nature of a supervisor's job is similar to that found in upper-level management positions but is in some respects more crucial. A supervisor is a critical link between the specialists—the professionals—and higher-level managers in the organization. A supervisor's subordinates are professional specialists who depend on receiving technical direction and general supervision from the supervisor. Professional subordinates see the supervisor performing many roles while supporting and working with them. A professional who assumes management responsibilities is soon confronted with many key questions concerning how to perform effectively as a manager. Such questions as the following are likely to arise in the new manager's mind:

- What knowledge, skills, and attitudes are essential to my success as a manager?
- How do I go about improving the performance of the professionals for whom I am responsible?
- How should I relate to my subordinates? Superiors? Peers?
- What must I do to provide a climate which will motivate people in their work?
- How do I relate to my former peer group?

[29] Paraphrased in part from Bertram N. Abramson and Robert Kennedy, *Managing Small Projects,* TRW Systems Group, TRW Inc., 1969.

- How can I get the work done by delegating it?
- What can be done to identify and stop problems before they get out of control?
- How do I maintain positive discipline?
- How can I be sure that I am communicating to the people who work for me?

Let's say that this is your first job as a manager. When you became a manager you started on a road to technical obsolescence! You were given a manager's job because you demonstrated technical competence in a professional field. You showed some common sense and an ability to work cooperatively and productively with other people. The personal and professional qualities that got you this far include the following:

- Demonstrated knowledge and skill in a specialized professional field
- Positive attitudes in support of organizational objectives and strategies
- A rapport with people
- Loyalty to and respect for persons in authority over you
- An ability to communicate in your technical specialty
- Pride in your technical specialty
- Self-confidence
- Working pretty much on your own on problems that have been assigned to you by a manager
- An image in the organization as a person who gets things done
- An ability to deal successfully with the challenge of doing quality work
- Evidence of a potential to be a leader in your specialty
- Assuming responsibility for your own actions

All of the above qualities are admirable and you would like to see them in the people that will work for you when you are appointed as a manager. But be careful! Some of them may not be important to you as a manager, and they may get in your way of performing effectively as a manager. What then are the personal attributes that you should seek to develop to be successful as a manager?

MANAGERIAL KNOWLEDGE, SKILLS, AND ATTITUDES[30]

The knowledge, skills, and attitudes that will enable you to survive and go far as a manager include

[30] Ibid.

- An ability to lead people along a way toward attainment of organizational objectives and goals.
- Enough knowledge and skills in a professional field to ask the right questions and to know if you are getting the right answers from the people you supervise in that field.
- Loyalty to the organization in which you are a manager.
- The trust of the people with whom you work—your subordinates, superiors, peers, and associates.
- Patience to let the people who work for you decide how the parts of a job are going to be done, even if you know how they ought to be done. As a manager you are more interested in results, not so much on how they are obtained.
- Predisposition to continue to build on a personal body of knowledge on the theory and practice of management.
- Conceptual powers to visualize the larger totality of a management problem or opportunity.
- A tolerance to watch someone else do a job that you may be able to do faster and better, and resist the temptation to do it yourself.
- Acceptance of the decline of your abilities in the specifics of a professional field as you spend more time as a manager than as a professional doer.
- Commitment to support, without reservation, the objectives and strategies of the organization to which your belong.
- Forebearance to spend as much time as necessary to do your job as a manager. You are no longer an eight-to-five person. Instead, you are going to use whatever time is required to deal with many people who help you in some way to do your management job.
- A competence in identifying problems and opportunities that need to be worked on. You can't wait for someone else to tell you what to do. You have to decide that by working with and observing other people.
- An interpersonal style in working with people such that they have confidence in your leadership abilities, fairness, and support of them in their work.
- Intellectual toughness to make management decisions affecting other people in the face of risk and uncertainty, and usually using inadequate information.
- Acceptance of the fact that there are no completely reliable laws or principles governing the behavior of people. The same input may produce different reactions from different people.
- A predisposition to select some courses of action based more on your intuition than a set of verifiable facts.
- Dependence on other people, for without their trust, loyalty, commitment, confidence, and support you will not succeed as a manager.

- A demonstrated competence in *followership*.
- An ability to perform many roles in your managerial capacity.

A professional who becomes a manager often faces a bewildering complex of forces and factors. A reorientation from technical things and concepts to people and abstractions is required. In technical work there are laws and principles that govern decisions. In management there is much artful style involved, just as there is the opportunity to use elements of science in the manager's job. A few suggestions are in order for managers-to-be to concentrate on in their managerial roles.

- Be patient—it takes time to become a successful manager, particularly in the development of artful skills in dealing with people and building personal relationships that are vital to gaining the respect and friendship of the people on whom you depend.
- Develop an appreciation of the science of management, that is, work towards the possession of a body of knowledge of management which covers the general truths or the operation of laws, principles, concepts, and techniques which can guide your thoughts and actions as a manager.
- Acquire the confidence to make and implement decisions in an environment that always contains elements of risk, uncertainty, and ambiguity. Be prepared to make a few mistakes, without which nothing creative or innovative is achieved.
- Never underestimate the ability of the people who work with you to read for themselves the implications of developments that are likely to affect their work. Get them involved, seek their help. They know more about how to do their job that you can ever possibly know.
- Personal leadership can motivate professionals to solve their problems. Work with them through counseling and talking through the things that are important to the solution of organizational and technical problems.
- Recognize that a person plays many roles when performing in a managerial capacity.[31]

[31] See H. Mintzberg, *The Nature of Managerial Work* (New York: Harper & Row 1973); and his article, "The Manager's Job: Folklore and Fact," *Harvard Business Review,* **53**(4) (1975):49–61. See also D. I. Cleland and D. F. Kocaoglu, *Engineering Management,* (New York: McGraw-Hill, 1981), Chapter 4.

ROLE PLAYING IN MANAGEMENT

A professional manager performs many roles in a managerial capacity. We suggest the following different yet related roles:

- A *leader* who leads people along a way to reach agreed-upon purposes
- An *entrepreneur/innovator* who helps to improve productivity in the organization
- A *subordinate* who works for another manager
- A *salesperson* who must become proficient at salesmanship in selling ideas to other people
- A *facilitator,* one who works with others to make their work easier, more rewarding, productive, and meaningful in contributing to organizational goals
- A *negotiator* for obtaining resources and maintaining a supportive environment so that people can work with economic, social, and psychological satisfaction
- A *specialist* in providing special know-how and leadership for the organization
- An *agent* who acts on behalf of other people
- A *logistician* who procures, distributes, maintains, and replaces the resources needed to do a job
- A *planner* who thinks through what resources are needed in the future to keep the organization going
- A *counselor* who listens and participates in an exchange of viewpoints and opinions on the nature of the work and offers advice and guidance to professionals on their performance
- A trusted *mentor* who instructs the subordinates on what results are expected of them and who enforces organizational strategy, policies, and procedures which keep the organization on the road to where it wants to go
- A *motivator* who helps to create the conditions that bring the best performance out of the people involved
- An *organizer* who organizes the resources required to do a job
- A *figurehead* who represents the organization in matters of formality
- A *decision maker* who chooses the alternative that best fits the organizational strategy
- A *teacher* concerned with instructing people on how resources are to be used to further organization's purpose

These roles are frequently carried out through working with teams.

ORGANIZATIONAL TEAMS

The importance of teams within any organization is self-evident from a review of how contemporary organizations are structured. Drucker states that five distinct structures emerge in modern organizations, one of which is the "team organization" which is set up for a specific task rather than a specific skill or stage in the work process.[32] In its original form the team was thought suitable only for short-lived, transitory assignments. Today it is used for permanent tasks in organizations. Teams demand self-discipline from each member; everyone works to do the team's work. Individuals cannot do their own "thing" on the team. Drucker concludes that organization builders will have to learn that sound organization structure needs both "(a) a hierarchical organization structure of authority, and (b) a capacity to organize task forces, teams, and individuals for work on both a permanent and a temporary basis.[33] For example, Mintzberg describes the use of teams as a form of "adhocracy" in which a fluid structure in which power is constantly shifting and coordination and control are by mutual adjustment worked through the informal communication and interaction of the experts in the organizations.[34]

Greiner maintains that growing organizations evolve through several phases of development. He suggests that the last phase includes spontaneity in management action through teams which focus on solving problems that cut across functions for task-group activity. A matrix organizational structure is frequently used to put together the right teams for the appropriate problems.[35]

There are several distinguishing features of the team:

- Two or more people have been identified to cooperate for the accomplishment of a distinct objective for the team as an entity, or the solution of a specific problem.
- The team is a suborganizational entity of a larger organizational totality.
- The team has been created as an outgrowth of an organizational strategy, policy, or action plan.
- Membership on the team may be full-time or part-time. Accordingly,

[32] Peter F. Drucker, "New Templates for Today's Organizations," *Harvard Business Review*, January–February, 1974, p. 51.
[33] Ibid., p. 53.
[34] Henry Mintzberg, "Organization Design: Fashion or Fit?" *Harvard Business Review*, January–February 1981, pp. 103–115.
[35] Larry E. Greiner, "Evolution and Revolution as Organizations Grow," *Harvard Business Review*, July–August 1972, pp. 37–46.

the members may have dual reporting relations: to the team and its leadership, as well as to their regular administrative supervisor.

- Responsible members of the team may feel dual alliances: to the team, as well as to their "home" organizational unit.
- The membership is usually drawn from different elements of a parent organization and/or separate organization.
- The functions of management are carried out within the team through the team leader and a consensus process within the team.
- The team may develop a distinctive culture through the interaction of the members and the influence of environment.
- The team is created to harmonize and integrate separate, and perhaps autonomous, organizational resources towards a common purpose to satisfy a larger organizational need.

HOW TEAM MEMBERSHIP BENEFITS THE INDIVIDUAL

Membership on a team provides the individual certain benefits–and responsibilities. In reviewing these benefits and responsibilities we will examine the influence that groups have on our lives. The reader is reminded that we are dealing with *formal* groups or teams in our discussions. Nevertheless the impact of *informal* groups on formal teams should be recognized.[36]

People, like many other animals, have a predilection to band together. Our first experiences in joining together come about through an affiliation with the family unit during infancy and youth. One could argue that we really don't have any choice in belonging to a group in these early years for sustenance, love, affection, self-esteem, security, and early experiences in appreciating a sense of belonging. Family and early childhood friends are the first groups which provide us with the satisfaction of affiliating and being accepted by others or, on the other hand, of being shunned by a group we would like to join. What child has not experienced the pain of being shunned by siblings or playmates? Children who have no social contact with siblings or playmates tend to find their life unsatis-

[36] In a broad sense groups are either *formal* or *informal*. *Formal* groups are those that exist as defined by the formal organizational structure of the enterprise, where people have assigned work responsibilities, tasks, activities, and team assignments. Behavior in the formal work group is influenced and supervised to contribute toward organizational objectives and goals. *Informal* groups are associations of people that are not structured nor designated by the management policies and procedures of the enterprise. Informal groups form at the initiation and desire of the constituent members who feel a need to join the group for social contact and other benefits. Both formal and informal organizations exist simultaneously in organizations.

fying. When children grow up and leave the home—and for the rest of their lives—the desire for social and professional interaction with a group is a persuasive need. Family and friends provide groups with which most of us can interact away from our jobs. At work we have a need for belonging to provide us with

- Identification
- Acceptance
- Interaction
- Communication
- Approval, empathy, and emotional support
- Performance norms
- Protection and assistance

Identification

Military commanders have reported situations where servicemen have performed heroic acts because of identification with their group. For example, a second lieutenant in Vietnam recommended for the Congressional Medal of Honor, made his attack against a fortified Viet-Cong machine gun nest because "what would the fellows have thought of me if I had been afraid to do it."[37] Soldiers in World War II who identified with cohesive groups were more effective in their responsibilities, less fearful in battle, and less likely to surrender under stress.[38] Other experiences indicate that heroic deeds were performed more because of identification and loyalty to a group with which the individual identified than anything else. It is the small group which individuals commit themselves to through identification.[39]

We tend to be joiners—affiliating with many different groups during our lifetime, starting out with neighborhood play groups, then joining groups in churches, schools, or such organizations as Cub Scouts and Brownies. We identify with these groups, and are usually proud of our affiliation with them. Many young people join the Marine Corps; anyone who has talked with a Marine will recognize how strongly the individual identifies

[37] Quoted in *The New York Times,* November, 17, 1966.

[38] E. A. Shils, "Primary Groups in the American Army," in Robert K. Merton and Paul F. Lazarsfeld (eds.), *Continuities in Social Research: Studies in the Scope and Method of the American Soldier* (New York: Free Press, 1950), pp. 16–39.

[39] S. Srivastva, S. L. Orbert, and E. H. Neilsen, "Organizational Analysis Through Group Processes: A Theoretical Perspective for Organizational Development," in Ian C. Cooper (ed.), *Organizational Development in the U.K. and the U.S.A.* (New York: Macmillan, 1977), pp. 83–111.

with the ideals and standards of the Marine Corps. In college, students join fraternities, sororities, "Independent Greeks," and other student organizations, in part to be identified with what is believed to be an elite group. Country clubs, United Airlines Red Carpet Club, the Sierra Club, all serve the purpose of providing the joiner with a sense of belonging through identification with a group.

Acceptance

Most of us feel the need to be accepted by other people. Acceptance in the family is important to the child. Acceptance of a loved one is a powerful force which holds families and friends together. The basis for acceptance ranges from a desire to speak a "common" language through the need for interacting, or to provide a basis for one's self-image. Acceptance is a social need, described by Maslow in the context of need for love, affection, and affiliation. The drive to belong to and be accepted by a group is described by Schein in the context of the experiences of the U.S. prisoners of war in North Korea.[40] There were repeated instances of collaboration with the enemy and few escape attempts. Schein suggests that the POWs' behavior was a result of the manner in which they were treated:

- Officers were separated from enlisted personnel, thus precluding any opportunity for formal organizations to be formed. (In contrast, during World War II, American officers and enlisted people held in German prisoner camps were for the most part integrated into strong formal organizations.)
- Groups that did form were systematically broken up; prisoners were transferred between barracks to reduce the chances of any strong informal groups being formed. (This probably accounted for the low escape attempts, with no organized escape attempts involving many prisoners. Without any organization the development of a plan for an escape was virtually impossible).
- Insufficient trust for escape developed among the prisoners, since there was no group identification and acceptance, prerequisites for the building of trust within any group.
- Morale was low in the camps because there was little interpersonal trust, no group solidarity, and certainly no sense of belonging through interaction with members of a group.

---

[40] Edgar Schein, "The Chinese Indoctrination Program for Prisoners of War," *Psychiatry*, **19** (May 1956):149–172.

The experiences in Vietnam, where the Army tried to provide its officers an opportunity to command in a combat operation, are believed by many to have been a management disaster. Through a rotation policy, officers served only six months in combat while the enlisted persons served 12 months. Morale fell so badly that more than 1000 officers were estimated to have been assassinated by their own troops. Rather than rotating whole units of Army personnel, the Army adopted a system whereby individuals instead of units were fed into the replacement system, a plan modeled after a system used in the auto industry to replace spare parts. Combat units because associations of strangers, and the rate of mutinies and combat refusals rose to the highest point in American history.[41]

Military historians have long recognized that the key to a successful combat unit is not so much the result of weaponry or even the quality of training of the people. Rather it is the result of strong bonds of shared attachment among the military members of the battle unit, which in turn is a function of sharing the same hardships, the same risks, the same common environment and fate, and the existence of a leader respected and accepted by the members of the military unit. Strong bonds in any group are only possible when the group has established strong bonds of alliances among the members and a clear leader is evident in the group.[42] In Vietnam the Army lacked a sense of uniqueness and group solidarity which had characterized military units in former times. As Richard A. Gamriel states: "Military organizations . . . require high levels of individual identification with institutional goals as the primary mechanism for compelling individual behavior. . . . In general the Army needs models of organizational development that stress personal ties, social interaction, stability of leadership elements and a sense of common identification. Such models are available in our society. One need only look at the success of monasteries, fraternal brotherhoods, religious groups and even cults as examples."[43]

The amount of collaboration with the enemy can be explained by the lack of group norms which would have assisted the individual prisoner to define the ethical behavior that was expected. In any group, the individual soon learns through interaction with the group's members what is desirable and unacceptable behavior.

---

[41] Richard A. Gabriel, "What the Army Learned from Business," *New York Times*, April 15, 1979.

[42] Ibid.

[43] Ibid.

Interaction and Communication

Individuals, for the most part, enjoy and seek interaction with others, particularly where there are common interests. We learn how to get along and survive at our place of work by interacting with knowledgeable people to learn the ropes in doing our job. Informal communication with members of the peer group helps to fill in where policy and standard operating procedures may be lacking. What are the generally accepted rules of the game? How far can one go in trusting the boss? Who is really boss in the product engineering department? These and similar questions cannot be answered by formal policies and procedures, but rather through interaction and communication with members of the appropriate group. An important function of interaction is to find out how acceptable behavior is defined in the group.

Behavioral Norms

All groups have norms, however informally defined, which dictate acceptable standards of behavior in the group. In the Hawthorne Studies it was found that norms reinforced worker output that was consistent with what the work group had determined to be a proper day's work. Work groups have norms which prescribe

- Dress codes
- Absenteeism
- Promptness or tardiness at the workplace
- Amount of horseplay allowed
- Fraternization codes
- Work quality
- Ethical and moral behavior
- Work output
- Behavior during work
- Street language

It is the small group which sets the standards for production and enforces member behavior to conform to these standards.[44]

Group pressures for conformity can influence an individual's judgment and attitude. Asch reported that when one person's opinion of objective

---

[44] F. J. Roethlisberger and W. J. Dickson, *Management and the Worker* (Cambridge, Mass.: Harvard University Press, 1939).

data was different from others, pressure was felt by the deviant to change opinion to conform with the others.[45] All of us have had the experience of "listening to a different drummer" and have felt the pressure to conform to what the rest of the people are doing. Our ability to withstand such pressures is greatly sustained if at least one other person joins us in nonconformity. Such numbers protect us and help us to get our views to prevail, or in some situations, to survive.

Protection and Assistance

Webber reminds us that "a fundamental function of the family is to protect its members from a hostile world."[46] We would add the following phrase to this statement: "And prepare them for survival and growth in a hostile world." Maslow's hierarchy of needs places safety and security needs at the second level, next to the basic physiological needs for sustenance and rest. Groups provide safety and protection as well as assistance in helping the individual get done what needs to be done. Co-workers help each other in doing their job. The "family feeling" within Delta Airlines conditions the way people in that organization help each other. During peak travel times pilots and administrators pitch in to help handle the baggage. Sometimes co-workers will cover up for each other in safety matters or during short absences from the production line. During a labor strike union members will go to great lengths to protect themselves from being persuaded by management to cross the picket line. Those who do leave the group of strikers and cross the picket line will be called scabs or something worse.

There is a positive side to how groups protect and assist each other. Training the newcomer in how to be successful at doing the work is an obvious benefit for all. In fact, most people when asked to help a newcomer will respond positively. Perceptive newcomers will try to identify some old-timer under whose mentorship the ropes of the organization can be learned. Those newcomers who are helped are usually expected to reciprocate by doing something for the "instructor" such as arranging for lunch or refreshments, or helping the instructor to finish up work. The ability to help out someone is often a source of satisfaction and prestige to the helper. If someone is approached properly for help within a group, it is rare for that help not to be forthcoming. When someone is helped to

---

[45] Solomon Asch, "Opinions and Social Pressure," in Harold J. Leavitt and Lewis R. Pondy (eds.), *Readings in Managerial Psychology* (Chicago: University of Chicago Press, 1964), pp. 304–314.

[46] Ross A. Webber, *Management* (Homewood, Ill.: Irwin, 1979), p. 101.

perform his duties better, there is a bonus in the form of emotional support.

EMOTIONAL SUPPORT

People join groups because of their need to be close to other people. Having people with whom you can bounce ideas around provides informal therapy and helps the individual to get a sense of how they are doing. Just having someone who will listen in a nonjudgmental and nonthreatening manner is useful. Parents will recognize how often children need someone to listen.

Status

Status is a position or rank of a person in relation to others. It may be bestowed by several means:[47]

- Association with an organization—"I'm with the foreign service."
- Occupation—"She's a medical doctor."
- Organizational level and job title—"He's a full professor."
- Work schedule—"Working all daylight."
- Salary level—"I make seventy-five thousand dollars a year."
- Mobility—"Moving around the plant to do her job."
- Awards—"He's a recipient of our outstanding teacher award."

Status may be acquired by informal means without an organizational sanction such as educational level, age, knowledge, skill, experience, personal attributes, or attitude. The old-timer who is willing to listen and passes his knowledge on to newcomers in the organization enjoys status from those newcomers. In our egalitarian society, many of us claim to eschew status yet we find ample evidence of status symbols in the form of large offices, impressive titles, high pay, distinctive or sophisticated clothing, prestigious automobiles, large homes, or social practices such as belonging to a distinctive club. A young engineer who is appointed as the manager of a small engineering project will feel a flush of status through being a member of the management group in the organization. A medical doctor obviously has status with the hospital staff where the patients are located; doctors also have status in their local community.

[47] Paraphrased from Stephen P. Robbins, *The Administrative Process* (Englewood Cliffs, N.J.: Prentice-Hall, 1976), pp. 284–285.

The importance of status can be appreciated by the results of a study reported by Kahn.[48] In a survey of 3000 workers from 16 industries, professional and white-collar occupations were reported as providing greater satisfaction than skilled tradesmen and blue-collar workers.[49] Robbins opines that this may explain the upgrading of titles: garbage collectors who become sanitary engineers, salesmen who are titled service engineers, or building handymen who are called superintendents.[50] The growing participation of the workers in matters heretofore reserved for managers has provided workers with increased status. Some managers perceive this as a loss of some of the power, status, and authority they once had. Some companies that have had separate dining facilities are consolidating eating facilities into one dining room available to all employees. For some people this represents a loss of status. Mellon Bank opened its officers' dining room, which had been available to only 1700 senior managers, to all employees. General Motors has eliminated private dining rooms at some plants as part of an overall plan to involve more workers in decisions.[51]

As the individual benefits by belonging to a team, so do the team and the team leader benefit; yet the team leader is dependent on how effectively the team performs.

TEAM DEPENDENCY

A manager's performance is inexorably tied to organizational unit (team, department, or higher level) for which the manager is responsible. A manager may carry out the functions of planning, organizing, motivating, directing, and controlling in an effective manner, yet if the organizational unit for which the manager is responsible does not attain its objectives, or attains the objectives in an inefficient manner, that manager has not been successful. Of course it is not this simple—but it does highlight the point that a manager's performance is interdependent with that of his subordinates, superiors, and peers. If an athletic team losses a game, all members of the team—the captain, coach, manager, and owners—lose.

Managerial performance depends a great deal on other people, many of whom may be somewhat distant, organizationally speaking, from the manager. Consider the example of a company that is trying to sell its products in a new market. To simplify the example, let's assume that this

---

[48] Robert L. Kahn, "The Work Module," *Psychology Today,* February 1973, p. 39.
[49] Ibid.
[50] Stephen P. Robbins, *The Administrative Process* (Englewood Cliffs, N.J.: Prentice-Hall, 1976), p. 286.
[51] "Labor Letter," *Wall Street Journal,* January 31, 1984, p. 1.

company is in the business of designing, developing, producing, and installing computer systems in the electrical power generation business. Salesmen, who have a strong technical background and knowledge of the market, are sent out to call on prospective customers. At this point, success or failure in this marketplace is almost totally dependent on the knowledge, skills, and attitudes of these salesmen. The president of the company and the other managers are very dependent on the ability of these salesmen to produce results. If the salesmen succeed, so will the managers, the president, and the organization itself. Managerial performance is mutually dependent on the other people—wherever located—who have some interest in the outcome of the organization. A manager may provide leadership in the planning for and use of resources in the organization and in helping to provide a productive environment for the people to work. A prudent manager will find ways of getting feedback—reports, meetings, observations, etc.—to determine if the resources are being used effectively and efficiently.

SUMMARY

In this chapter we have presented an overview of team management by describing the situations where teams are used in contemporary organizations. We defined the character of management and described the professional's transition to managership. Finally we discussed how membership in a team benefits and exacts responsibilities from the individual.

In Chapter 2 we will discuss the nature of communication and its role in the management of teams.

BIBLIOGRAPHY

Baday, M. K. "Easing the Transition to Management." *Machine Design,* June 11, 1981.

Boettinger, Henry. "Is Management Really an Art?" *Harvard Business Review,* January–February 1975.

"Breaking with Tradition." *Data Corporation Third Quarterly Report,* September, 30, 1982.

Cleland, David I. "Matrix Management (Part II): A Kaleidoscope of Organizational Systems." *Management Review,* December 1981.

Cleland, David I., and Dundar F. Kocaoglu. *Engineering Management.* New York: McGraw-Hill, 1981.

"Data General's Management Trouble." *Business Week,* February 9, 1981.

Drucker, Peter F. "New Templates for Today's Organizations." *Harvard Business, Review,* January–February, 1974, p. 51.

Magnet, Myron. "Managing by Mystique at Tandem Computers." *Fortune,* June 28, 1982.

Main, Jeremy. "Anatomy of an Auto-Plant Rescue." *Fortune,* April 4, 1983.

Mee, John F. "Matrix Organization." *Business Horizons,* Summer 1964.

Mintzberg, H. *The Nature of Managerial Work.* New York: Harper & Row, 1973.

Mintzberg, H. "The Manager's Job: Folklore and Fact." *Harvard Business Review,* **53**(4) (1975).
Pascale, Richard Tanner. "Zen and the Art of Management." *Harvard Business Review,* March–April, 1978.
Peters, Tom, and Nancy Austin. "A Passion for Excellence." *Fortune,* May 13, 1985.
Scobel, Donald N. "Doing Away with the Factory Blues," *Harvard Business Review,* November–December 1978.
Serrin, William. "The Way That Works at Lincoln." *New York Times,* January 15, 1984.
Williamson, Merritt A., P. E. "Engineering Schools Should Teach Management Skills." *Professional Engineer.*

2
COMMUNICATIONS

THE NATURE OF COMMUNICATION

Communication is the process by which information is exchanged between one or more individuals. Since the management process thrives on communication, it is extremely important that new or prospective managers be thoroughly familiar with how managers communicate. Unfortunately, even the best-run organizations may be plagued by communication problems, at all levels of management.

Communication is the lifeblood of organizations. Effective communications includes talking, listening, writing, and reading. In fact, *any* means that an individual uses to convey information, feelings, attitudes, or simply thoughts involves the communication process. The communication process may be formal, informal, written, or oral. Lower-level employees (i.e., technical ranks) may spend 50% or more of their time communicating, whereas the upper-level personnel may have a significantly higher percentage.

Nichols surveyed white-collar workers and found that the average employee was communicating 70% of the time.[1] The forms of communication involved were listening, 45%; speaking, 30%; reading, 16%; and writing, 9%.

Communication can be verbal or nonverbal. Nonverbal communication may include the way we dress, comb our hair, or wear makeup. Facial expressions such as smiling or frowning are also examples of nonverbal communication. Such examples convey a message (i.e., information) to a culturally oriented observer.

In order to convey the message correctly, it is important to understand not only *what* you wish to say, but *how* to say it. There exists a growing trend today for written language to follow spoken language patterns, thus creating a greater chance that the information can be misinterpreted. Effective communication requires an understanding of the blockages of effective communications, when they occur, and how to eliminate them.

[1] Ralph G. Nichols, "Listening Is Good Business," *Management of Personnel Quarterly*, Winter 1962, p. 2.

COMMUNICATION COMPLEXITIES

No concept is more basic for effective project management than communication. Good project communication allows the project manager to get the message across with the meaning which was intended, whereas poor communication leads to ineffective management, poor leadership, mistrust, and low project morale. Although communication is not magical or mystical, the process must be continuously nurtured, cultivated, and modified by the project manager.

The complex nature of the communication process can best be understood when we realize that there does not exist any one universal definition of communication. The following definitions are often found in the literature:[2]

- An exchange of information
- An act or instance of transmitting information
- A verbal or written message
- A process by which meanings are exchanged between individuals through a common system of symbols
- A technique for expressing ideas effectively

There are two major reasons why the project communication process is complex. First, there are several media by which information can be exchanged such as reports, memos, team meetings, personal conversations, or telephone. Second, the receiver can be an individual at any position of an organization or simply a personal friend. This is shown in Figure 2-1. The complexity arises because the project manager must match the proper media with the individual who will receive the message. Project managers are responsible for multidirectional communications, most frequently with individuals with whom they may never have interfaced before. This is where the true problem lies because, if a situation of communications mistrust occurs, then the result will be an increased requirement for documentation, more meetings, possibly on-site representation, and potential project termination or project manager dismissal.

The growth of modern business has created enormous requirements for a continuous flow of information. Unfortunately, even with giant computers which can store massive data bases and are capable for microsecond retrieval, we are still buried with extraneous information and must therefore establish some type of filtration process, time- and cost-effectiveness such that the required knowledge is readily available.

---

[2] Source unknown.

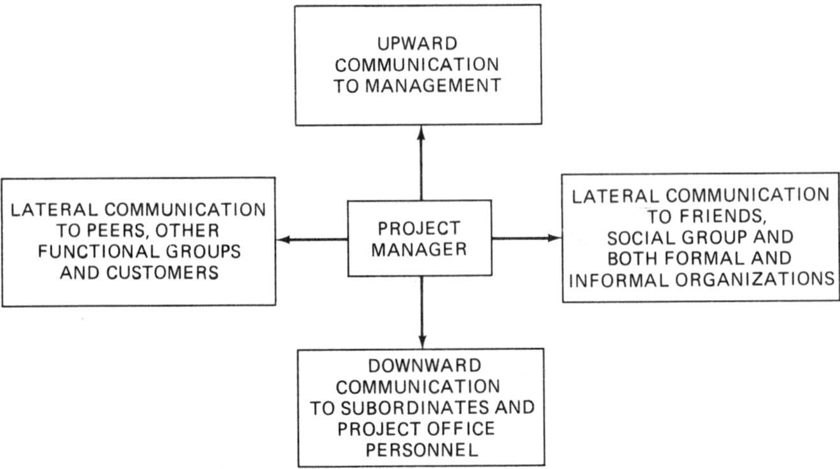

Figure 2-1. Communication channels.

Naisbitt has commented that "we are drowning in information but starving for knowledge." According to Naisbitt,

> this level of information is clearly impossible to handle by present means. Uncontrolled and unorganized information is no longer a resource in an information society. Instead, it becomes the enemy of the information worker. Scientists who are overwhelmed with technical data complain of information pollution and charge that it takes less time to do an experiment than to find out whether or not it has already been done.
>
> Information technology brings order to the chaos of information pollution and therefore gives value to data that would otherwise be useless. If users—through information utilities—can locate the information they need, they will pay for it. The emphasis of the whole information society shifts, then, from supply to selection.[3]

Effective project communication is needed to ensure that we get the *right* information to the *right* person at the *right* time and *in a cost-effective manner*. As organizations grow and mature, the need for more information becomes apparent. The difficulty is in determining what project information is critical and should be transmitted. The cost of obtain-

[3] John Naisbitt, *Megatrends* (New York: Warner Books, 1982), p. 24.

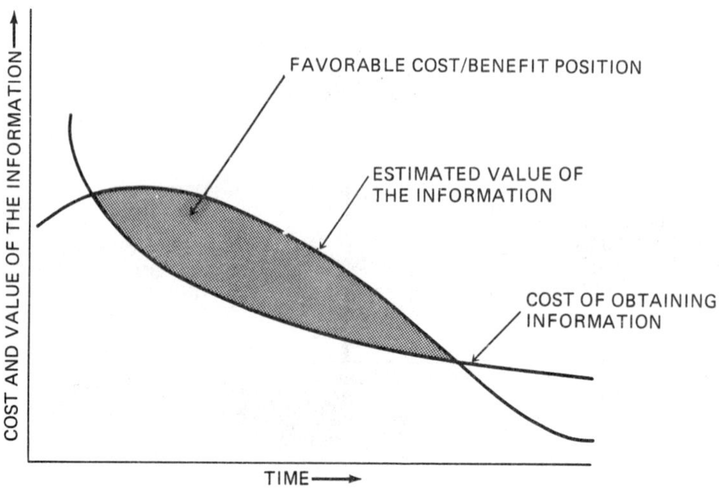

Figure 2-2. Cost/benefit analysis. *Source:* Harold Kerzner, *Project Management: A systems approach to planning, scheduling and controlling* (New York: Van Nostrand Reinhold, 1979), p. 29.)

ing this information must be less than the value of this information in order for benefits to be realized. This is seen in Figure 2-2. The shaded in area of Figure 2-2 shows that region where it is beneficial to the organization to pay for the additional information. Outside of this region, the value of the information is not worth the cost of obtaining the information, unless for the sake of developing better customer relations or good will.

Most companies today are attempting to develop management cost and control systems. The difficulty is in determining what input/output information is needed to accurately determine project status. Another problem is in determining the cost-effectiveness of obtaining this information (i.e., How often should schedules be updated? What is the cost of updating? How accurate are the percent completion inputs?).

Communication complexities have caused management to reassess the training needed by new or prospective engineering project managers. In a recent survey, personnel were asked: "If you had the opportunity to broaden your communication skills, in which of the following areas would you seek training?"[4] Respondees were asked to rank the areas in order of need. The results are shown in Table 2-1.

[4] Source unknown.

Table 2-1. Survey Results.

TOPIC	RANKED BY MANAGEMENT RESPONDENTS	RANKED BY ENGINEERING RESPONDENTS
Rapid reading	1	1
Public speaking	2	4
Conference leadership	3	7
Report writing	4	2
Interviewing skills	5	10
Listening skills	6	3
Oral presentation (papers)	7	5
Engineering graphics	8	9
Composition and rhetoric	9	8
Letter writing	10	6

COMMUNICATION NETWORKS

Organizations cannot function effectively unless their members are able to communicate effectively. This fact has long been known by organizational development (O.D.) personnel, organizational planners, and organizational behavioralists who contend that organizational structures should be designed so that the resulting communication networks provide a free flow of information throughout the organization. The military chain of command is often used as an example of centralized information flow, whereas a matrix structure functions on multidirectional or decentralized information flow (even though a project manager exists!).

Figure 2-3 shows various communication networks used in experimental investigations. The circles represent positions, the lines represent communications channels, and the arrows indicate one-way channels. The networks shown in Figure 2-3 do not have to be congruent with the organizational structure. For example, the informal organization within a company might use a completely different communication network than the formal structure. Experienced project managers know how to use both types of organizations effectively.

Communication networks have been investigated to see the effect that patterns of communication have upon organizational development. For example, project management organizational structures such as a matrix cannot readily adapt to ever-changing external environment without unrestricted multidirectional communication.

People who occupy communication focal point positions generally emerge as leaders. Again using the matrix structure as an example, pro-

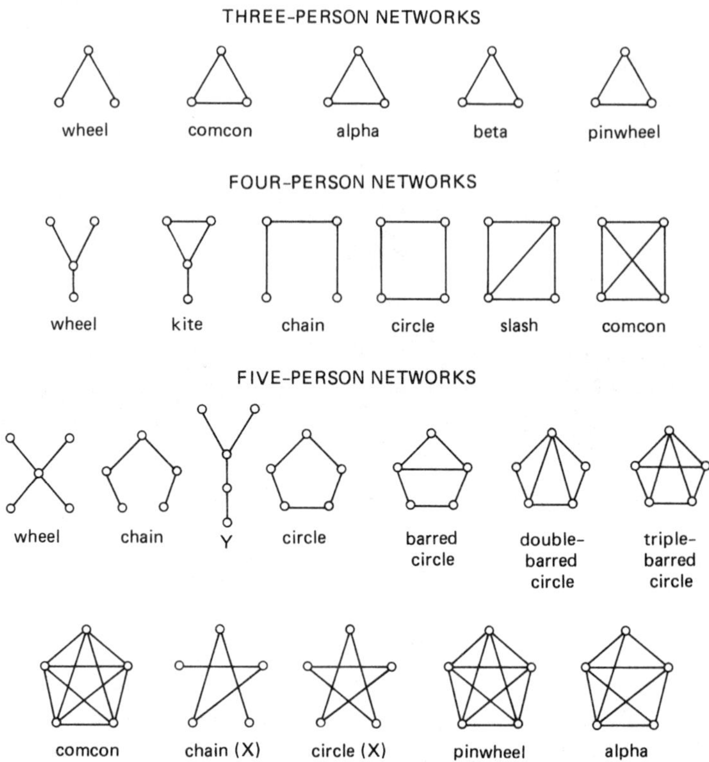

Figure 2-3. Communication networks used in experimental investigations. Circles represent positions, lines represent communication channels, and arrows indicate one-way channels. (Reprinted with permission from M. E. Shaw, "Communication networks," In L. Berkowitz (ed.), *Advances in Experimental Social Psychology,* Vol. 1 (New York: Academic Press, 1964), pp. 111–147.)

ject managers are required to have strong communicative skills and act as the focal position in the organization for all project communication. The success (as well as failure) of the project and the project manager might very well rest with the project manager's communicative skills. Today, contracts are not necessarily awarded to the lowest bidder, but moreso to those companies that have a previous history of effective management *coupled with effective customer-contractor relations.*

In organizations which thrive on project management, the major criteria for advancement through the project management ranks may very well be the project manager's ability to communicate, whereas in the classical or vertical organizations, promotion may be based upon a different set of criteria such as technical competence. Almost all organizations reluctantly admit that they have managers (in key positions) who cannot communicate effectively because promotions were based upon technical rather than managerial merit. As a result, the communication networks must either be rebuilt, destroyed, or modified. It is truly a shame what can happen to a well-functioning communication network with the promotion of a weak manager.

People who sit in focal positions in communication networks generally have more job satisfaction than those individuals who sit in the peripheral positions.

Unfortunately, too many people believe that the way to get ahead in the company is to become a communication focal point where power and authority are directly related to one's ability to control the flow of information. Such individuals in centralized positions have the power to overload or saturate a communication network to the point where efficiency and effectiveness deteriorate.

Gilchrist et al. were the first to identify the concept of saturation.[5] According to these authors, there are two kinds of saturation: channel saturation and message saturation. Channel saturation refers to the number of channels with which a communication position must deal, whereas message saturation refers to the number of messages handled by the communication position. Some networks are more susceptible to saturation than others. As an example, the five-person "wheel" network of Figure 2-3 is more vulnerable to saturation than other network models.

THE COMMUNICATION MODEL

The process by which information is exchanged between individuals can be shown by the model in Figure 2-4. In the simplest terms, the sender (i.e., the project manager) transmits a message to a receiver. Communication is more than simply conveying a message; it is also a *source for control.* Proper communications inform employees because employees need to *know* and *understand.* This may very well be the project manager's key to successful motivation of personnel. Communication must convey both information and motivation. The problem, therefore, is how to communicate. Below are six simple steps:

1. Think through what you wish to accomplish.
2. Determine the way to communicate.

[5] J. C. Gilchrist, M.E. Shaw, and L. C. Walker, "Some Effects of Unequal Distribution of Information in a Wheel Group Structure," *Journal of Abnormal and Social Psychology,* **49** (1954):554–556.

Figure 2-4. Communication process.

3. Appeal to the interest of those affected.
4. Give playback on ways others can communicate to you.
5. Get playback on what you communicate.
6. Test effectiveness through reliance on others to carry out your instructions.

Unfortunately, the sender must encode the message such that the receiver understands the meaning, and the receiver must decode the information and interpret what the sender was trying to convey. The elements of the communication model include the following:

- *The source/encoder* is the person who is originating the message. Encoding is the process of putting the information into a form in which can be received and understood by another. There are several variables which affect the source as the information is encoded:[6]

    - Communication goals
    - Communication skills
    - Frame of reference
    - Sender credibility
    - Needs
    - Personality and interests
    - Interpersonal sensitivity
    - Attitude, emotion, and self-interest
    - Position and status
    - Assumptions (about receivers)
    - Existing relationships with receivers

- *The message* is the output of the encoding process and is the information which the source wished to transmit. The message can be either verbal or nonverbal. Verbal messages are language. Nonverbal messages include eye contact, facial expressions, posture, tone of voice, and mode of dress. Nonverbal messages can either enforce or reject verbal messages, and can be extremely important when the project

[6] Adapted from Howard M. Carlisle, *Management: Concepts and Situations* (SRA Publishers, 1976), p. 527.

manager must deal with new employees or ones with whom he has not come into contact previously.

- *The receiver/decoder* is the recipient of the message. Decoding is converting the message into some type of format where the information is understandable to the receiver, and hopefully with the same meaning as intended by the source. Decoding also includes interpretation variables such as[7]

  - Evaluative tendency
  - Preconceived ideas
  - Communications skills
  - Frame of reference
  - Needs
  - Personality and interest
  - Attitudes, emotion, and self-interest
  - Position and status
  - Assumptions about sender
  - Existing relationship with sender
  - Lack of responsive feedback
  - Selective listening

The decoding process is often regarded to be more difficult than the encoding process because individuals can speak at a rate of 100–200 words/minute whereas the brain functions much faster. As a result, the brain has more idle time during the communication process and evaluates other information which can distract from the original message.

- *The medium* is the carrier of the information and includes face-to-face communications, telephone conversations, reports, and group discussions. The selection of the medium depends upon such factors as speed of response, confidentiality, and formality.
- *Noise* is any factor that can influence the communication process. Noise may simply be distortion during the coding or encoding process, static, or the cluttering of communication channels with multiple messages. Noise has a severe impact on our field of experience. Internally to the individual, noise causes perception and interpretation barriers, whereas explicitly, environmental noise may cause the message to become garbled. An example would be giving an employee instructions while he is preoccupied with another activity. Noise tends to distort or destroy the information within the message.

---

[7] Ibid., pp. 527–528.

Noise results from personality screens that dictate the way we present the message, and perception screens that may cause us to "perceive" what we thought was said. Noise can therefore cause ambiguity.

- Ambiguity causes us to hear what we want to hear.
- Ambiguity causes us to hear what the group wants.
- Ambiguity causes us to relate to past experiences without being discriminatory.

Figure 2-5 shows the complete communication model. The screens or barriers are from one's perception, personality, attitudes, emotions, and prejudices, as defined previously under encoders and decoders.

- *Perception barriers* occur because individuals can view the same message in different ways. Factors influencing perception include the individual's level of education and region of experience. Perception problems can be minimized by using words that have precise meaning.
- Personality and interests, such as the likes and dislikes of individuals, affect communications. People tend to listen carefully to topics of interest but turn a deaf ear to unfamiliar or boring topics.
- Attitudes, emotions, and prejudices warp our sense of interpretation. Individuals who are fearful or have strong love or hate emotions will tend to protect themselves by distorting the communication process. Strong emotions rob individuals of their ability to comprehend.

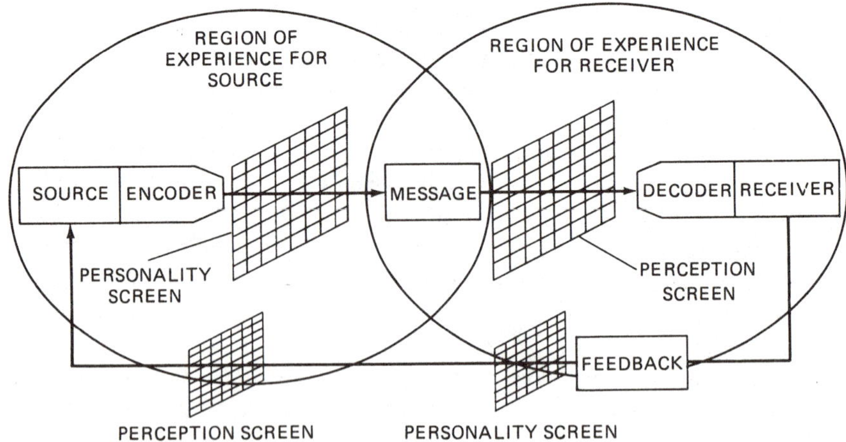

Figure 2-5. Total communication process.

With every effort to communicate, there will be barriers. The most common barriers include

- Hearing what they want to hear. This results from people doing the same job so long that they no longer listen to the project manager.
- Sender and receiver having different perceptions. This is vitally important in interpreting contractual requirements, statements of work, and proposal information requests.
- Receiver evaluating the source before accepting the communications.
- Receiver ignoring conflicting information and doing as he pleases.
- Jargon meaning different things to different people.
- Ignoring nonverbal cues.
- Receiver being upset emotionally.

The scalar chain of command can also become a barrier with regard to in-house communications. The project manager must have the authority to go to the general manager or his counterpart to communicate effectively. Without direct upward communication, it is possible that filters can develop so that the final message gets distorted.

Figure 2-5 also identifies the feedback channel. Feedback is the response of the receiver to the transmitted message. The feedback channel is a verification process to ensure that the receiver correctly decoded the transmission from the source. During verbal communications, the sender often asks the receiver to repeat, in his own words, his interpretation of the communication.

If the regions of experience overlap, as they do in Figure 2-5, then the sender and receiver have a *common base* from which to encode/decode the message. This is one of the reasons why project managers try to staff projects with personnel with whom they have worked previously. However, if the regions do not overlap, then there is probably a good chance that the message will be misinterpreted. Overlapping regions tend to reduce the size of the ambiguity barriers, thus ensuring a greater likelihood of successful communication. This is one of the reasons why companies are leaning toward life-cycle project planning with standardized planning documents.

Three important conclusions can be drawn from communications techniques and barriers:

- Don't assume that the message you sent will be received in the form you sent it.
- The swiftest and most effective communications take place among people with common points of view. The project manager who fos-

ters a good relationship with associates has less difficulty in communicating with them.
- Communications must be established early in the project in order to have sufficient time to obtain feedback.

MISINTERPRETING THE MESSAGE

No matter how hard we try to communicate clearly, information can easily be misinterpreted. The development of sophisticated policies and procedures does not always alleviate such problems. Consider the following situations:

- A company priced out a large contract which called for the development of a new product. The contract stated that a *minimum* of 15 tests be conducted. The company priced out 20 tests just as a safety precaution. The customer felt that the results of the first 15 tests were inadequate and demanded that an additional 15 tests be conducted, causing a cost overrun of $40,000. Management has since decreed that contracts of this nature identify a maximum as well as a minimum testing quantity.
- A computer company priced out a Navy aerospace contract that required the prototype to be tested in *water*. The prototype, once completed, was submerged to the bottom of a swimming pool for testing. Unfortunately, the Navy's definition of the word "water" was deep-sea Atlantic Ocean. This resulted in a cost overrun of $1,000,000 because the computer company had to send their test equipment and test engineers on an Atlantic voyage.

These two situations identify the danger of poorly planned communication and illustrate the following points:

- Even the simplest words or expressions can be misinterpreted under the right situation.
- Written communications offer more opportunity for misinterpretation because instantaneous feedback may not be possible.
- Written communications should allow sufficient time for the meaning of the message to be verified.

OVERCOMING COMMUNICATION BARRIERS OF THE SENDER

The majority of all communication problems in organizations are attributed to a lack of understanding and awareness of barriers. When the effects of these barriers are understood, corrective action should be possible.

Cheryl L. McKenzie and Carol J. Qazi have identified a simple method for overcoming barriers:

> When it becomes evident that something has gone wrong in a conversation, some steps can be taken to re-establish communication and get the job done:
>
> - Try to recall the gestures that accompanied the verbal message. Restate the message without the gestures.
> - Remember what was actually stated and what other directions or requests wee implied in that statement. State all messages verbally.
> - Recall what words and phrases were used in making the request or giving the directions. If they were idiomatic, restate using more formal language.
> - Recall the manner in which the message was delivered. Repeat the message more slowly.[8]

The following suggestions should help to improve organizational communication for the source/encoder:

- Communication goals should be set prior to message dissemination. Setting goals includes defining exactly what the message is to accomplish. Good communicators always seek to clarify not only their ideas but the true purpose of the communication. Consulting with others may also be appropriate.
- Appropriate verbal and nonverbal language should be used. This requires that the total setting (physical and human) be considered.
- The source/encoder must be *empathetic* with the problems that the receiver will have in interpreting the message.
- The source/encoder can drastically improve his credibility by the necessary actions to support the message. Simply stated, don't make promises that cannot be kept. Message effectiveness increases when the message contains information of help or value to the receiver.
- Effective communication requires that a trusting climate exist between the source/encoder and receiver/decoder. This makes the follow-up or feedback process easier to perform.
- Managers must *select the appropriate* media for message transmission.

---

[8] Cheryl L. McKenzie and Carol J. Qazi, "Communication Barriers in the Workplace," *Business Horizons,* March–April 1983, p. 72.

The American Management Associations has identified "Ten Commandments of Good Communications:

- Seek to clarify your ideas before communicating. The more systematically we analyze the problem or idea to be communicated, the clearer it becomes. . . . Good [communication] planning must [also] consider the goals and attitudes of those who will receive the communication and those who will be affected by it.
- Examine the true purpose of each communication. Before you communicate, ask yourself what you really want to accomplish with your message—obtain information, initiate action, change another person's attitudes? Identify your most important goal and then adapt your language, tone, and total approach to serve that specific objective.
- Consider the total physical and human setting whenever you communicate. Meaning and intent are conveyed by more than words alone. . . . Consider for example, your sense of timing—i.e., the circumstances under which you make an announcement or render a decision; the physical setting—whether you communicate in private, for example, or otherwise; the social climate that pervades work relationships within the company or a department and sets the tone of its communications; custom and past practice—the degree to which your communication conforms to, or departs from, the expectations of your audience. . . .
- Consult with others, where appropriate, in planning communications. . . . Such consultation often helps to lend additional insight and objectively to your message. Moreover, those who have helped your plan your communication will give it their active support.
- Be mindful, while you communicate, of the overtones as well as the basic content of your message. Your tone of voice, your expressions, your apparent receptiveness to the responses of others—all have tremendous impact on those you wish to reach. Frequently overlooked, these subtleties of communication often affect a listener's reaction to a message even more than its basic content. . . .
- Take the opportunity, when it arises, to convey something of help or value to the receiver. Consideration of the other person's interests and needs—the habit of trying to look at things from his point of view—will frequently point up opportunities to convey something of immediate benefit or long-range value to him.
- Follow up your communication. This you can do by asking questions, by encouraging the receiver to express his reactions, by follow-up

contacts, by subsequent review of performance. Make certain that every important communication has a "feedback" so that complete understanding and appropriate action result.

- Communicate for tomorrow as well as today. While communications may be aimed primarily at meeting the demands of an immediate situation, they must be planned with the past in mind if they are to maintain consistency in the receiver's view; but, most important of all, they must be consistent with long-range interest and goals. For example, it is not easy to communicate frankly on such matters as poor performance or the shortcomings of a loyal subordinate—but postponing disagreeable communications makes them more difficult in the long run and is actually unfair to your subordinates and your company.
- Be sure your actions support your communications. In the final analysis, the most persuasive kind of communication is not what you say but what you do. . . . For every manager this means that good supervisory practices—such as clear assignment of responsibility and authority, fair rewards for effort, and sound policy enforcement—serve to communicate more than all the gifts of oratory.
- Seek not only to be understood but to understand—be a good listener. When we start talking we often cease to listen—in that larger sense of being attuned to the other person's unspoken reactions and attitudes. . . . [Listening] demands that we concentrate not only on the explicit meanings another person is expressing, but on the implicit meanings, unspoken words, and undertones that may be far more significant.[9]

OVERCOMING COMMUNICATION BARRIERS OF THE RECEIVER: EFFECTIVE LISTENING

Communication barriers for receiver/decoder can be overcome by understanding the principles of effective listening. Business, today, has been so burdened with communications that emphasis is being placed upon oral rather than written messages. Unfortunately, the effectiveness of oral communications hinges more upon how people listen than how they talk. As Sperry Corporation has stated in advertisements, "looking for the solution without listening to the problem is like working in the dark."

---

[9] Adapted, by permission of the publisher, from "Ten Commandments of Good Communication," *Management Review*, October 1955 © 1955 American Management Association, Inc. All rights reserved.

Consider the following excerpts:[10]

- "Listening on the average is at a 25% level of efficiency."
- "The fact is, there's a lot more listening than hearing."
- "After we hear something, we must interpret it. Evaluate it. And finally respond to it. That's listening."
- "We prejudge—sometimes even disregard—a speaker based on delivery or appearance."
- "We let personal ideas, emotions or prejudices distort what a person has to say."
- "We tune-out subjects we consider too difficult or uninteresting."
- "And because the brain works faster than most people speak, we too often wander into distraction."
- "Yet as difficult as listening really is, it's the one communication skill we're never really taught."

These excerpts can be supported by comments made to Nichols and Stevens:[11]

- "Frankly, I had never thought of listening as an important subject by itself. But now that I am aware of it, I think that perhaps 80% of my work depends on my listening to someone, or on someone else listening to me."
- "I've been thinking back about things that have gone wrong over the past couple of years, and I suddenly realized that many of the troubles have resulted from someone not hearing something, or getting it in a distorted way."
- "It's interesting to me that we have considered so many facets of communication in the company, but have inadvertently overlooked listening. I've about decided that it's the most important link in the company's communications, and it's obviously also the weakest one."

Effective listening requires developing empathy with the speaker, controlling or eliminating distractions, avoiding arguments and criticism, monitoring your own listening habits, and keeping an open mind. Keith Davis has prepared a list of ten commandments for good listening:

---

[10] Excerpts taken from "Knowing How to Listen Takes More Than Two Good Ears," *Wall Street Journal,* September 15, 1980, p. 11.

[11] Ralph G. Nichols and Leonard A. Stevens, "Listening to People," *Harvard Business Review,* September–October 1957, p. 112.

- Stop talking! You cannot listen if you are talking. Polonius (Hamlet): "Give every man thine ear, but few thy voice."
- Put the talker at ease. Help the talker feel free to talk. This is often called a permissive environment.
- Show the talker that you want to listen. Look and act interested. Do not read your mail while he or she talks. Listen to understand rather than to oppose.
- Remove distractions. Don't doodle, tap, or shuffle papers. Will it be quieter if you shut the door?
- Empathize with the talker. Try to put yourself in the talker's place so that you can see his or her point of view.
- Be patient. Allow plenty of time. Do not interrupt the talker. Don't start for the door or walk away.
- Hold your temper. An angry person gets the wrong meaning from words.
- Go easy on argument and criticism. This puts the talker on the defensive. He or she may "clam up" or get angry. Do not argue: even if you win, you lose.
- Ask questions. This encourages the talker and shows you are listening. It helps to develop points further.
- Stop talking! This is first and last, because all other commandments depend on it. You just can't do a good listening job while you are talking.[12]

Nature gave us two ears but only one tongue, which is a gentle hint that we should listen more than we talk.

ORGANIZATIONAL COMMUNICATION

Generally speaking, there are three types of communication: downward, upward, and lateral, as shown in Figure 2-1. Each type will be discussed in more detail in later sections. Organizational communication is any information that generally follows the organizational chart and chain of command. Such information includes policies, procedures, rules, guidelines, and directives.

Upward communication is either feedback, evaluative, or approval-type information. Feedback up the organization is necessary so that upper-level managers will know what is actually happening in the organiza-

[12] From *Human Behavior at Work* by Keith Davis. Copyright © 1972 by McGraw-Hill Inc. Used with permission of McGraw-Hill Book Company.

tion. Such information includes technical and status reporting, customer information, or simply documents requiring approvals.

Lateral communication is information that flows horizontally across the organization. This is perhaps the hardest type to achieve because line groups generally tend to compete with one another rather than cooperate. As a result, lateral communication often requires formal authority to accompany it. Such is the case in various forms of Project management which require that work flow horizontally throughout the organization such that *all* personnel are working toward the same goal and objective.

Executives contend that managers in the twenty-first century will have such strong communicative skills that the span of control can be increased in most organizations. This may create a problem, as depicted in Figure 2-6. As the span of control increases, the organizational chart will appear as an inverted T. The management positions included in the shaded sections will then no longer be needed. What should senior management do with these managers that are no longer needed as managers?

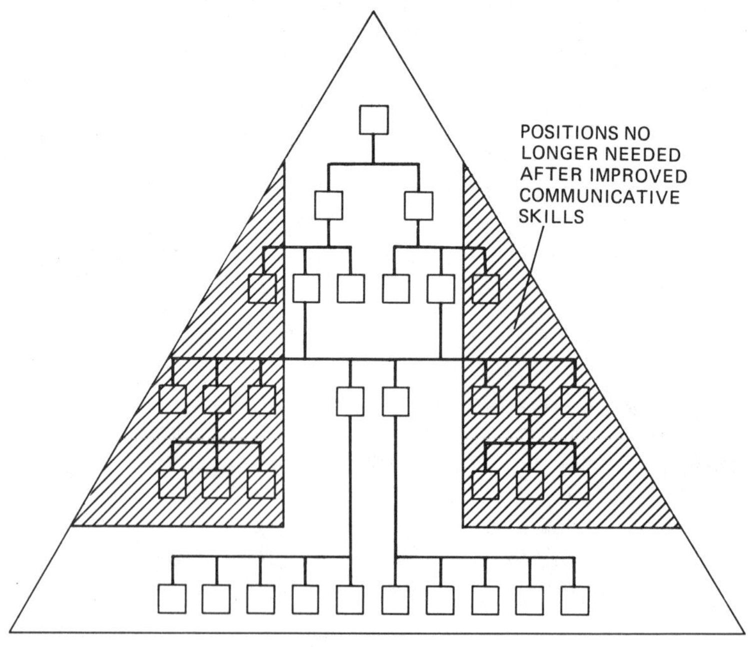

Figure 2-6. Management span of control.

PATTERNS OF ORGANIZATIONAL COMMUNICATION

There exist both formal and informal patterns of communication. The most common type of formal communication is the *serial transmission,* which involves passing on information from one individual to another. Of course the apparent weakness of such serial transmissions is that messages become distorted as the length of the transmission increases. Distortions occur as a result of information being added, deleted, or altered. A well-known article by Alex Bavelas and Dermot Barrett illustrates three patterns of (serial transmission) communication and the influence of the group characteristics of speed, accuracy, organization, emergence of a leader, and morale. The patterns studied are shown in Figure 2-7.

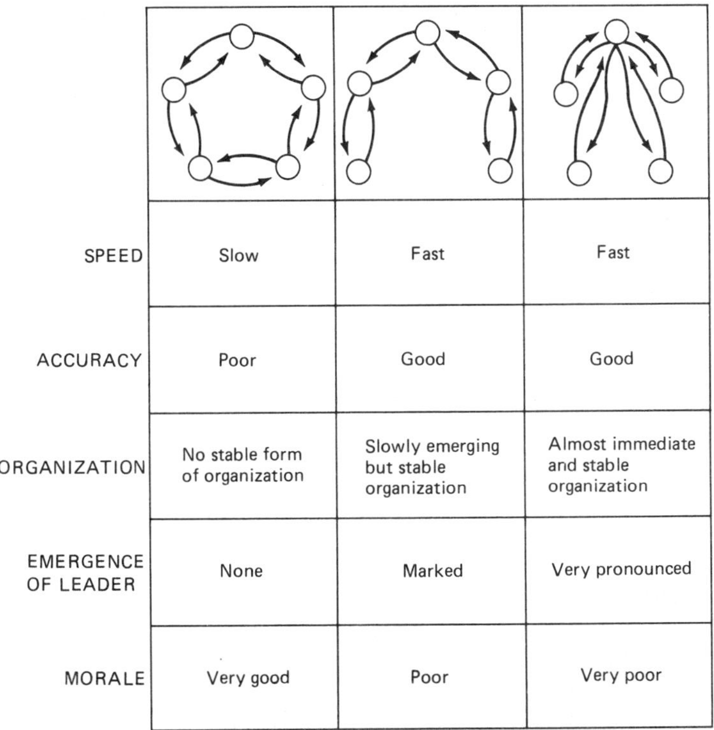

SPEED	Slow	Fast	Fast
ACCURACY	Poor	Good	Good
ORGANIZATION	No stable form of organization	Slowly emerging but stable organization	Almost immediate and stable organization
EMERGENCE OF LEADER	None	Marked	Very pronounced
MORALE	Very good	Poor	Very poor

Figure 2-7. Relationship between three patterns of organizational communication and group characteristics of speed, accuracy, organization, emergence of leader, and morale (Reprinted, by permission of the publisher, from "An Experimental Approach to Organizational Communication" by Alex Bavelas and Dermot Barrett, *Personnel,* March 1951, © 1951 by American Management Association, Inc., p. 370, All rights reserved.)

Individuals placed in the communication focal positions often become extremely powerful, yet perhaps have very little authority. The following remarks by Art Buchwald illustrate this point.

> The thing that hit me when I first come back to the United States was how powerful the American secretary has become in the American way of life.
>
> I hadn't realized it until I tried to make an appointment with a successful college chum who was pulling down around $40,000 a year in a big company (which shall remain nameless as his secretary would never forgive him for speaking about her to me).
>
> After several attempts, I finally managed to get a luncheon date and he apologized profusely for the difficulty I had experienced.
>
> "You don't understand what's going on in the United States," he told me, looking around to make sure no one was listening. "The secretaries are taking over. No one can get through to me if my secretary decides she doesn't want him to."
>
> "She makes all my appointments, she decides when I can take a vacation, if it's safe for me to make a speech in another town. She watches me constantly and I swear I'm scared silly of her."
>
> "Why don't you fire her?" I asked him.
>
> He looked at me incredulously. "You must be out of your mind. You can't fire your secretary. She knows where all the bodies are buried. She's my espionage agent to let me know what's going on in the company.
>
> "Without the information she picks up from the other secretaries I wouldn't be able to last a week with the company. Besides, frankly I don't understand what I'm supposed to be doing in the company, and she does."
>
> "I can see your point." I said, watching him drink his third martini.
>
> He stared into the glass. "The only thing is, I wish she wouldn't hate my wife so."
>
> "Does she hate your wife?"
>
> "All the secretaries hate their bosses' wives," he said. "I don't think that it has anything to do with jealousy. It's just that secretaries think wives are so damn inefficient. They also feel that wives take up too much of their bosses' time. My secretary thinks that I could do a much better job if I didn't have to go home to my wife for dinner. And she believes my weekends with my family are a complete waste of time. She doesn't see how I can live with a woman who doesn't understand the company.

"Also, since my secretary pays all the bills, she thinks my wife is sort of a spendthrift. But to be honest, I'm so browbeaten by my secretary during the day, with her constant nagging and efficiency, that I really look forward to going home to my wife at night. I look on my wife as sort of a mistress, the only one who understands me."

"What does your wife think about your secretary?"

"She's afraid of my secretary. My wife has to be nice to her because if she isn't, my secretary won't let my wife speak to me. As it is, my secretary only lets her get through 50 percent of the time. The other 50 percent she just says I'm in an important meeting, as if to imply my wife should know better than to call the office when world-shaking events are going on behind the company's locked doors."

"I didn't realize secretaries were that powerful," I said sympathetically.

"You don't know the half of it. Look, if your secretary catches a cold and is out two days, you might as well shoot yourself. But if your wife catches pneumonia, all you have to do is come to the office and tell your secretary to notify the Blue Cross."[13]

Informal organizational communication or "grapevine" is a type of communication based upon friendship and relationships rather than one's organizational charter position. The informal organization generally exists to handle information not readily available to the individual through the formal channels of communication. Certo has identified three distinctive characteristics of grapevines:[14]

- The grapevine springs up and is used irregularly within the organization.
- The grapevine is not controlled by top executives, who may not even be able to influence it.
- The grapevine is used largely to serve the self-interests of the people within it.

The importance of the grapevine cannot be overstated. Suppose that top management made a change in the way the organization would function. Through what means of communication would the employees first

[13] Art Buchwald, *Reign of Terror: The Secretary's the Dictator*. © 1984 by The Los Angeles Times Syndicate.

[14] Samuel C. Certo, *Principles of Modern Management* (Dubuque, Iowa: Brown, 1980), p. 312.

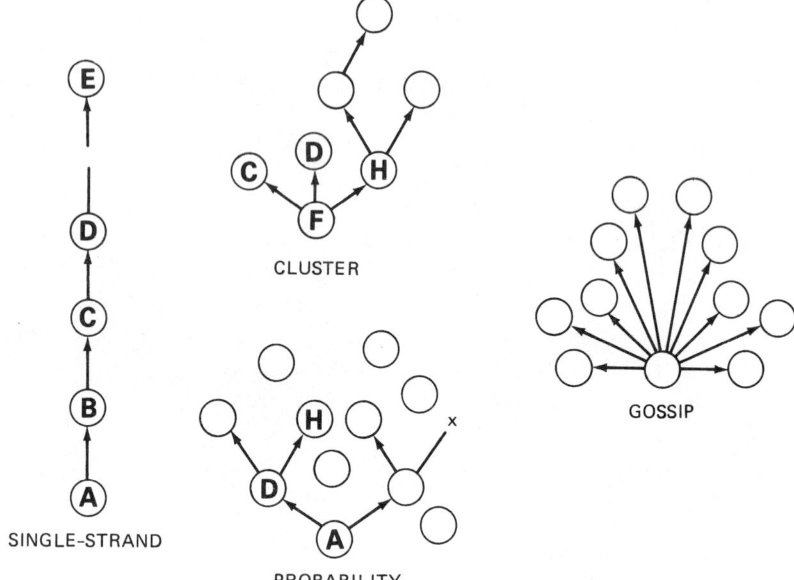

Figure 2-8. Four types of organizational grapevines. (Reprinted by permission of the *Harvard Business Review,* from "Management Communication and the Grapevine" by Keith Davis (September–October 1953), Copyright © 1953 by the President and Fellows of Harvard College, all rights reserved.)

get the word? According to Walton, the response to this question appeared as follows:[15]

- Grapevine, 38%
- Supervisor, 27%
- Official memo, 17%
- Station newspaper, 7%
- Station directive system, 4%
- Bulletin boards, 4%
- Other, 3%

Informal communications are serial transmissions and therefore prone to distortion, moreso than formal communications. Figure 2-8 identifies four grapevine patterns:

[15] Eugene Walton, "Communicating Down the Line: How They Really Get the Word," *Personnel,* July–August 1959, pp. 78–82.

- The single-strand grapevine is pure serial transmission where A tells B, who tells C, who tells, D, and so forth. Messages here have the greatest chance for distortion because each person may have a different personality and perception screen.
- The gossip grapevine has one person informing everyone else on the grapevine. This might be the executive's secretary informing all other grapevine secretaries concerning a change to take place.
- The probability grapevine is random communication where A informs D, who informs H, and so forth. In the probability grapevine, there is no guarantee that everyone will get the word because the information spreads on a random basis.
- The cluster grapevine is actually smaller grapevines within a larger grapevine. Here, A informs C, D, and F who, in turn, inform the members of their grapevines.

As shown in Figure 2-9, organizational communication forms the nervous system of an organization by tying together planning, organizing, staffing, controlling, and directing. Executives, if given the choice, would prefer formal rather than informal communication with the argument that grapevine communications are often nonfactual rumors that inflict more damage than good for the organization.

However, there are many executives that have found ways to cultivate the informal channels of communication to such a degree that beneficial

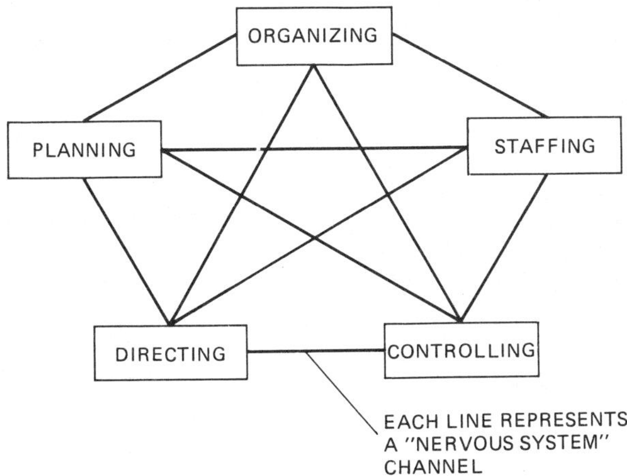

Figure 2-9. Management principles communication network: the nervous system.

results were obtained. Peters and Waterman, in their best-seller, *In Search of Excellence,* have made the following observations:

> The nature and uses of communication in the excellent companies are remarkably different from those of their nonexcellent peers. The excellent companies are a vast network of informal, open communications. The patterns and intensity cultivate the right people's getting into contact with each other, regularly, and the chaotic/anarchic properties of the system are kept well under control simply because of the regularity of contact and its nature (e.g., peer versus peer in quasi-competitive situations).
>
> The intensity of communications is unmistakable in the excellent companies. It usually starts with an insistence on informality. At Walt Disney Productions, for instance, everyone from the president on down wears a name tag with only his or her first name on it. HP is equally emphatic about first names. Then come the open door policies. IBM devotes a tremendous amount of time and energy to them. The open door was a vital part of the original Watson philosophy, and it remains in force today—with 350,000 employees. The chairman continues to answer all complaints that come in to him from any employee. Open door use is pervasive at Delta Airlines as well; at Levi Strauss it means so much that they call the open door the "fifth freedom."
>
> Getting management out of the office is another contributor to informal exchanges. At United Airlines, Ed Carlson labeled it "Visible Management" and "MBWA—Management by Walking About." HP Treats MBWA (Management by Wandering Around" in this instance) as a major tenet of the all-important "HP Way."
>
> Another vital spur to informal communication is the deployment of simple physical configurations. Corning Glass installed escalators (rather than elevators) in its new engineering building to increase the chance of face-to-face contact. 3M sponsors clubs for any groups of a dozen or so employees for the sole purpose of increasing the probability of stray problem-solving sessions at lunch time and in general. A Citibank officer noted that in one department the age-old operations-versus-lending-officer split was solved when everybody in the group moved to the same floor with their desks intermingled.
>
> What does it add up to? Lots of communication. All of HP's golden rules have to do with communicating more. Even the social and physical settings at HP foster it: you can't wander around long in the Palo Alto facilities without seeing lots of people sitting together in rooms with blackboards, working casually on problems. Any one of those ad hoc meetings is likely to include people from R&D, manufacturing,

engineering, marketing, and sales. That's in marked contrast to most large companies we've worked with, where the managers and analysts never meet or talk to customers, never meet or talk to salesmen, and never look at or touch the product (and the word "never" is not chosen lightly.) A friend at HP, talking about the company's central lab organization, adds: "We aren't really sure what structure is best. All we know for certain is that we start with a remarkably high degree of informal communication, which is the key. We have to preserve that at all costs." 3M's beliefs are similar, which led one of its executives to say, "There's only one thing wrong with your excellent company analysis. You need a ninth principle—communications. We just plain talk to each other a lot without a lot of paper or formal rigmarole." All of these examples add up to a virtual technology of keeping in touch, keeping in constant informal contact.[16]

COMMUNICATION WITH EXECUTIVES

The major function of vertical communication to executives is to keep senior management informed about what is happening in the lower levels of the organization. This type of communication is based upon the following needs and expectations of lower-level personnel:

- Clearly defined decision channels
- Support for decision making
- Progress reporting
- Feedback, explanations, suggestions, advice
- Support in communicating with other organizations
- Protection from political infighting
- Opportunities for growth

If the upward communication to executives must go through middle management, it is likely that the information will be filtered. There are several reasons for this filtration process. First, middle managers are likely to withhold information that may not reflect favorably upon middle management. Second, middle managers view their job as protecting upper management from nonessential data, thus altering, filtering, or shortening the message. Third, employees or middle managers who have high aspirations for upper-level mobility or visibility are less likely to communicate information that can be interpreted in a detrimental fashion concerning

[16] Thomas J. Peters and Robert H. Waterman, Jr., *In Search of Excellence* (New York: Harper & Row, 1982), pp. 121–123, 262.

their ability or performance. Fourth, individuals tend to filter out any bad news if they feel that senior management has the power to punish them in some way. Last, employees tend to withhold information if a situation of mistrust occurs between employees and executives. According to Stoner:

> Even unambitious subordinates will be guarded in their communications if an atmosphere of trust does not exist between them and their superiors. Subordinates conceal or distort information if they feel that their superiors cannot be trusted to be fair or that they may use the information against them. The net result of all these communication problems is that higher-level managers frequently make decisions based on faulty or inadequate information.[17]

Executive communication must be short and to the point. Executives are interested more in end results than in tools, techniques, or means to an end. One executive commented that any report to upper management that contains a paper clip or staple is too long.

COMMUNICATING WITH SUBORDINATES

Communication downward to subordinates is most always filtered as managers decide exactly what and how much information should be passed down. It is often the case that managers transmit either partially complete or inaccurate information in order to keep subordinates dependent upon them. The net result is that employees become uninformed concerning policies and procedure changes, and there is a lack of meaningful data. A situation of mistrust and lack of company loyalty soon reigns as employees may be unable to carry out their tasks properly.

Downward communication is generally regarded as more difficult than upward communication because of the number of perception and personality screens that exist among subordinates. As a result, subordinate feedback loops are extremely important and most first-level supervisors utilize a two-way communication process. Leavitt and Mueller have identified advantages and disadvantages of one-way and two-way communication.[18]

- One-way communication takes considerably less time than two-way communication.

[17] James A. Stoner, *Management* (Englewood Cliffs, N.J.: Prentice-Hall, 1982), p. 510.
[18] Harold J. Leavitt and Ronald A. H. Mueller, "Some Effects of Feedback on Communicating," *Human Relations*, **4**(4) (November 1951):401–410.

- Two-way communication is more accurate than one-way communication. (That is, the diagrams were more accurately reproduced when two-way communication was used.) The feedback allows the sender to refine his or her communication for the receivers so that it becomes more precise and accurate.
- Receivers are more sure of themselves and of their judgments when two-way communication is used. The very fact that they are permitted to ask questions probably increases the receivers' self-confidence. In addition, they can use questions to clarify any doubts they may have.
- Senders can easily feel attacked when two-way communication is used, because receivers will call attention to the senders' ambiguities and mistakes.
- Although it is less accurate, one-way communication appears much more orderly than two-way communication, which often appears noisy and chaotic.

The most common method for subordinate communication is the KISS rule: *k*eep *i*t *s*imple, *s*tupid. The manager-to-subordinate communication can be written or oral, whereas the feedback from subordinate to superior is oral. Asking a subordinate to repeat the instructions in his own words is an effective way of seeing if the employee really understands the communication.

CUSTOMER COMMUNICATIONS

Customer communications play a vital role in the growth and survivability of all organizations. Today, contracts are not necessarily awarded to the lowest bidder, but perhaps to the second or third lowest bidders based upon management capability and previously established good customer-contractor communication channels.

Quite often, the communication channels between a customer and contractor are based upon equal salary levels, status, or prestige. Figure 2-10 shows such channels for a project-driven subcontractor. The executive and employee horizontal channels are informal, whereas the project manager's horizontal information flow is formal.

In Figure 2-10 the executive's responsibility is to provide executive client contact. At the senior management level, executives prefer to communicate with executives rather than lower-level personnel. Thus it becomes extremely important for project managers to provide executives with timely information concerning project status. The primary purpose for the executive-to-executive communication channel is to convince the

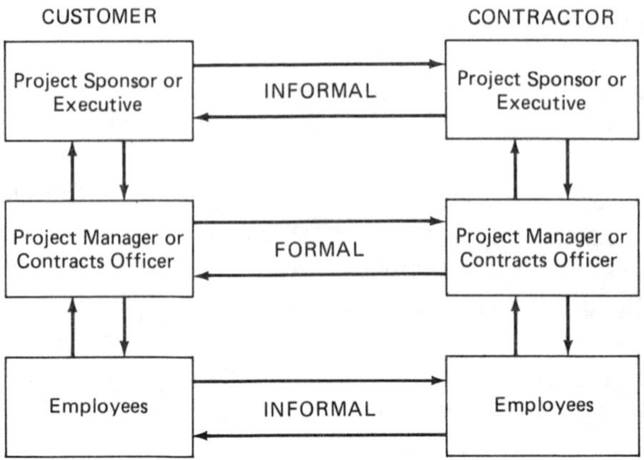

Figure 2-10. Customer communications.

customer that there is an executive in the contractor's organization who is sincerely interested in how the customer's money is being spent. The second purpose is to provide checks and balances for the formal channels of communication, that is, to verify that the correct information has been exchanged between customer and contractor.

The most dangerous level of communication is between the contractor's and client's employees themselves. This type of communication is very hard to stop because the customer believes that all information flowing through the formal channel of communication has been filtered and therefore he wants his employees to be able to communicate one-on-one so as to obtain first-hand information. There are two problem areas with this type of communication. First, the employees may see only a small portion of the total picture and their remarks may be construed as a company (rather than individual) opinion. Second, employees may try to make commitments for a change in the scope of the project, additional work, or testing, none of which are possible without authorization by the formal channel of communication. Verification of employee-to-employee communication must be made in writing, with sign-off from both the customer and the contractor.

THE CULTURAL SIDE OF COMMUNICATIONS

The growth of American business today seems to be more multinationally oriented than domestically oriented. This orientation poses severe cultural problems for many corporations because many of the communication variables are not known. As the cultural communication variables and barriers increase, misunderstandings will increase.

Managers involved in multinational efforts will have to undergo increased training in communications, social customs, policies, and procedures. Harris and Moran have identified four cross-cultural situations that affect communication and understanding between two culturally different individuals.

> You are involved in a technical training program in Iran and one of your responsibilities is to rate persons under your supervision. You have socialized on several occasions and you and your family have spent some time with one of the Iranians you are supervising. The Iranian is an extremely friendly and hardworking individual, but had difficulty exercising the leadership expected of him. On the rating form you so indicated; and this was discussed with him by his supervisor. Subsequently, he came to you and asked how you could have said that about him because you were friends. You indicated that you also thought you were friends, but that you had an obligation to report deficiencies and areas to be improved upon to his supervisors. This statement was incomprehensible to the Iranian. . . . What cultural differences might cause misperceptions of performance appraisal and evaluation?
>
> You are negotiating a contract with a Japanese company and during the meetings there are times of silence on the part of the Japanese negotiating team. The negotiations, from your perspective as an American, appear to be proceeding at an unusually slow pace and even the simplest decisions of commitments appear to take an inordinate amount of time. You begin to push a little harder and your frustrations mount as you begin to hear statements from the Japanese such as, "it will take a little more time" and "this is quite difficult." What should we do? What is happening to the negotiating process? How do the Japanese negotiate?
>
> You are in Saudi Arabia attempting to finalize a contract with a group of Saudi businessmen. You are aware these people are inveterate negotiators, however, you find it difficult to maintain eye contact with your hosts during conversations. Further, their increasing physical proximity to you is becoming more uncomfortable. You also have noticed that a strong handgrip while shaking hands is not returned. When invited to a banquet, because you are left-handed, you use your left hand while eating. Your negotiations are not successfully concluded. What may have been the reason for this? What cultural aspects are evidenced in

this interaction, which if known, could improve your communications with your Arab clients?

You are a black American manager of a group of Puerto Rican workers in a New York factory. You resent the constant use of Spanish among your subordinates because you only speak English. You suspect your employees use their language as a means of criticizing and mocking you—they are often laughing and you wonder what it is all about. In bicultural/bilingual areas with large Hispanic populations, why should supervisors have some knowledge of Spanish? Why do your subordinates feel more comfortable in their native language? How could your company facilitate their instruction in the English language? Or should it?[19]

CONDUCTING MEETINGS

Meetings provide an excellent method for information exchange together with immediate feedback. Some organizations thrive on meetings to such an extent that some managers believe that they simply go from meeting to meeting eight hours each day. The following list shows typical meetings of a construction company:[20]

COMMUNICATIONS

- Weekly staff meeting
- Telephone conferences with contractor
- Change control meeting
- Interface working groups
- Project management reviews
- Management councils
- Design reviews
- Outside review
- Quarterly management meeting (emphasis on cost)
- Contractor's performance reviews
- Frequent visits to contractor and subcontractor facilities
- Involvement of senior management
- Open-door policy

---

[19] Philip R. Harris and Robert T. Moran, *Managing Cultural Differences* (Houston: Gulf Publishing), pp. 20–21.
[20] Source unknown.

Meetings should be held whenever information must be transmitted. Unfortunately, too many meetings are scheduled simply because a certain time period has elapsed. Many meetings should not be held at all. Today, many managers are asking themselves what would happen if they didn't attend the meeting at all. "Meetingitis" has become a way of life in many companies.

Anthony Jay had identified classes of meetings:

> Meetings can be graded by size into three broad categories: (1) the assembly—100 or more people who are expected to do little more than listen to the main speaker or speakers; (2) the council—40 or 50 people who are basically there to listen to the main speaker or speakers but who can come in with questions or comments and who may be asked to contribute something on their own account; and (3) the committee—up to 10 (or at the most 12) people, all of whom more or less speak on an equal footing under the guidance and control of a chairman.[21]

Meetings can also be classified according to their frequency of occurrence:[22]

- The daily meeting, where people work together on the same project with a common objective and reach decisions informally by general agreement.
- The weekly or monthly meeting where members work on different but parallel projects and where there is a certain competitive element and a greater likelihood that the chairman will make the final decision himself.
- The irregular, occasional, or "special project" meeting, composed of people whose normal work does not bring them into contact and whose work has little or no relationship to the others'. They are united only by the project the meeting exists to promote and motivated by the desire that the project succeed. Though actual voting is uncommon, every member effectively has a veto.

Meetings must be effective or else they become time management pitfalls. It is the responsibility of managers and meeting chairmen to ensure that meetings are valuable and necessary for the exchange of information.

---

[21] Anthony Jay, "How to Run a Meeting," *Harvard Business Review*, March–April 1976, p. 123.

[22] Ibid., p. 124.

The following are general guides for conducting a more effective meeting:

- Start on time. If you wait for people, you reward tardy behavior.
- Develop agenda "objectives." Generate a list and proceed. Avoid getting hung up on the order of topics.
- Conduct one piece of business at a time.
- Allow each member to contribute. Support, challenge, and counter. View differences as helpful. Dig for reasons or views.
- Silence does not always mean agreement. Seek opinions: "What's your opinion on this, Betty?"
- Be ready to confront the verbal member: "Okay, we've heard from Mike on this matter, now how about some other views?"
- Test for readiness to make a decision.
- Make the decision.
- Test for commitment to the decision.
- Assign roles and responsibilities (only after decision making).
- Agree on follow-up or accountability dates.
- Indicate the next step for this group.
- Set the time and place for the next meeting.
- End on time.
- Was the meeting necessary?

Meetings are, for the most part, not spontaneous and must be prepared for. First, attendance must be limited or else the organization may suffer from a loss of productive time among key employees. Second, meetings should be no longer than an hour and a half to two hours or else the meeting loses its effectiveness. Third, the agenda of the meeting should be distributed in advance so that participants can review the data and make meaningful contributions. However, care must be taken not to send out the agenda too far in advance. Fourth, the seating arrangement must be considered. Classroom-style seating is ideal for reports and presentations, whereas circular seating provides information exchange but may encourage conflicts. Sometimes, encouraging a clash of ideas brings forth additional, productive information.

Handouts and viewgraphs should be prepared at least one week prior to the meeting. Meetings are more effective is participants are allowed to review the material in advance. Customers, in particular, do not like surprises in the middle of meetings and therefore require the review of all material before the meeting.

- "In any given meeting, when all is said and done, 90 percent will be said—10 percent will be done."—Orben's Current Comedy
- "A committee meeting provides a great chance for some people who like to hear their own voices talk and talk, while others draw crocodiles or lady's legs. It also prevents the men who can think and make quick decisions from doing so."—Lin Yutang, The Pleasures of a Nonconformist (World).
- "Having served on various committees, I have drawn up a list of rules: Never arrive on time or you will be stamped a beginner. Don't say anything until the meeting is half over; this stamps you as being wise. Be as vague as possible; this prevents irritating the others. When in doubt, suggest that a subcommittee be appointed. Be the first to move for adjournment; this will make you popular—it's what everyone is waiting for."—Harry Chapman, quoted in *Think*.

Many project managers do not understand when and how to use meetings effectively. Meetings are used to

- Communicate status to all employees in a reasonable manner
- Provide a common basis for decision making

Team meeting communication is a two-way street. The project manager provides proposal status information to the team members and seeks out necessary functional information. As an example, team meetings are a fast, effective means of updating and revising schedules and providing the information to all other employees on an equal basis.

Decision making poses a problem for the project manager because not all team members may have the authority to make decisions. Team meetings delegate authority to commit functional resources and make decisions. In this case, the project manager should send team meeting agenda items to the functional managers, together with a list of potential problems and decisions. This should indicate to the functional managers that someone with authority to commit resources is required to attend the team meeting. It is always a good policy to send out meeting agendas and an approximate time that each item will be discussed. This usually works well because line managers do not have the time to sit through a two-hour team meeting simply to discuss a 15-minute agenda item. If employees do not have the right to make decisions, then the manager may find that his team will flood the agenda with irrelevant items simply to avoid decision making. Finally, managers must be aware that it is possible to conduct too few team meetings as well as too many team meetings. The meeting chairman may also find the following problems:

- People tend to resist the exploration of new ideas.
- People tend to protect themselves.
- People tend to look at day-to-day activities rather than long-term efforts
- People tend to create win-or-lose positions.
- People look for individual rather than group recognition.
- Because of superior-subordinate relationships (i.e., pecking orders), creativity is inhibited.
- Criticism and ridicule have a tendency to inhibit spontaneity.
- Pecking orders, unless adequately controlled, can inhibit team work and problem solving.
- All seemingly crazy or unconventional ideas are ridiculed and eventually discarded. Contributors do not wish to contribute anything further.
- Many lower-level people, who could have good ideas to contribute, feel inferior and therefore refuse to contribute.
- Meetings are dominated by upper-level management personnel.
- The meetings are held at inappropriate places and times.
- Many people are not given adequate notification of the meeting time and subject matter.

WRITING REPORTS AND MEMOS

Effective management requires an exchange of information. Unfortunately, too many managers view the word *exchange* as meaning written rather than oral. The net result is that many managers feel trapped because of the increase in paperwork. Report writing is extremely time-consuming and costs much more than the average person realizes. A typical report requires.

- Writing
- Typing
- Proofing
- Editing
- Retyping
- Graphic arts
- Reproduction

It is not uncommon for corporations to spend between $100 and $200 per page of report. Many projects eventually overrun in the later stages because companies did not know how to price out the reporting requirements. Even with word processors, the costs are high.

There are three purposes for report writing. The first purpose is to seek action by making recommendations or to guide others responsible for policy and procedure development or implementation. The second purpose is simply to provide information on what is happening throughout the organization. The third purpose is to provide a record of work accomplished so that the information can be retrieved, if necessary, some time in the future.

There are three types of written media used in organizations:

- Individually oriented media: These include letters, memos, and reports.
- Legally oriented media: These include contracts, agreements, proposals, policies, directives, guidelines, and procedures.
- Organizationally oriented media: These include manuals, forms, and brochures.

Report preparation involves the logical sequence of researching the details, analyzing the details, and preparing the conclusions. Report writing is in reverse order, namely, writing the conclusions, the analyses of the details, and finally the details themselves. The methodology behind this approach is that the reader can start at the beginning of the report and continue on through to as much depth as he wants.

The actual writing of the report is a matter of one's own style and perhaps upbringing. Engineering companies maintain a staff of technical writers to provide some degree of consistency in report writing.

The following checklist can be utilized in evaluating reports:[23]

- On organization
  - Is the title clear? Is it accurate?
  - Is the summary or overview statement clear? Does it emphasize the really important ideas?
  - Is all the information in some logical sequence? Or do the ideas jump around?
  - Are the more important ideas ahead of the less important ones?
  - Is the report complete? If not, what is missing?
  - Overall, is the organization excellent, good, fair, weak, or poor?

[23] Source unknown.

- On clarity
  - Are any words more difficult than necessary?
  - Are the ideas divided properly into sentences of reasonable length?
  - Are there wasted words that can easily be deleted?
  - Do connectives provide logical flow between ideas?
  - Do passive-voice verbs withhold useful information?
  - Is the tone courteous?
  - Overall, is the clarity excellent, good, fair, weak, or poor?

The following guides can be used for preparing a memo or letter:

- Do we have only one objective?
- Is the objective stated in the first sentence?
- Is the memo/letter written as we talk (i.e., using action verbs and avoiding verbosity and formality)?
- Do we use personal pronouns and names?
- Is the average sentence limited to 15–20 words?
- Is the correspondence limited to one page?
- Did the correspondence end with an impact?
- Does the correspondence contain eye appeal, margins, indentations, capitals, short paragraphs, type style, and quality paper?

Most people explain their thoughts clearly in person but find it difficult to do the same in writing. Writing should be done with impact, using exact language. Writing should be direct, concise, and clear so that your directions can be followed. The objective, of course, is to minimize the reader's work. The writer must think the same as the reader would think. Therefore, the writer must understand the psychology of the reader and write with confidence.

DELIVERING A SPEECH

Whether making a speech or a presentation, the objective is either to inform, convince, stimulate, or cause action to occur. To achieve these objectives, a well-planned preparation is necessary. The preparation process must include

- Knowing the age, sex, race, and background of the audience
- Knowing the mood of the audience

- Knowing exactly how much time you have and how much time, if any, should be allocated to questions and answers

The delivery of the speech should follow a logical pattern. It is an acceptable process to speak from an outline and to pause occasionally to review your material. A common method of avoiding this is through the use of transparencies. Transparencies or overheads are better to use because the room must be darkened to use slides.

There are three common methods to begin your speech. The first method involves storytelling or humor. The story should arouse interest and provide a smooth transition into your speech. Humor need not have the people rolling in the aisles to be effective. A simple smile or chuckle will suffice.

A second method is paying an honest compliment to the person introducing you or to the company/organization sponsoring the program. The third method is referencing your topic. This technique works well if your audience has strong feelings toward the subject.

Your conclusion to the speech should be short and sweet, perhaps with a brief review. You may wish to express your thanks to your audience, the organization, and your host.

INTERVIEWING

When actual or potential problems occur in an organization, management often resorts to interviewing to obtain the desired information. The most common types of interviews are telephone, questionnaire, and face-to-face. Telephone and questionnaire interviews have several limitations. First, people being interviewed are more likely to provide honest, meaningful data when they are acquainted with the interviewer or simply are face to face with the interviewer. On the other hand, some employees are more open with interviewers whom they do not know, believing that they will not be punished for what they say.

During face-to-face interviewing, the following guidelines can be used:

- Create an informal, relaxed environment.
- Explain the reason for the interview.
- Take notes rather than tape the interview.
- If management agrees, offer the interviewees the opportunity to see the conclusions when available.
- Be honest with the interviewees.

COMMUNICATION PHILOSOPHIES

There are basically four types of philosophies which people use when communicating.[24] Most people utilize more than one philosophy, depending upon the situation.

The *controlling communicator* believes in one-way communication. He recognizes himself as the expert and therefore does not require any feedback. There are no alternatives to his ideas. He generally tries to sell or force his ideas upon subordinates.

Many executives adopt this philosophy, thinking that they must have done something right to get where they are. As managers move higher and higher up the chain of command, they tend to become more autocratic and their communication philosophy approaches that of the communicator.

There are situations in which a controlling philosophy is acceptable. These situations include

- R&D projects where the project leader is recognized as the
- In time of crisis or when the project is approaching time, cost expert, and performance constraints.
- When speed is of the utmost importance (and where accuracy can be waived

The *developmental communicator* believes in the systems approach to management and communications, and employs a "we" approach to decision making. The developmental communicator encourages new ideas, recommendations, and alternatives. Since this philosophy encourages two-way communication, speed may have to be sacrificed.

The *relinquishing communicator* believes that other individuals are more knowledgeable than he is and therefore he has no confidence in his own abilities. He understands effective management and communications, and is often willing to explore alternatives, but wants others to make the final recommendations.

A relinquishing communicator can work well with a controlling communicator, but will probably never achieve a higher (or even equal) level than the controlling communicator. Controlling communicators often avoid relinquishing authority and responsibility.

Good managers often use the relinquishing philosophy together with other philosophies. The relinquishing philosophy works well if someone

[24] These philosophies have been adapted from Malcolm E. Shaw, *Developing Communication Skills* (West Point, Conn.: Educational Systems and Designs, Inc., 1968), pp. 30–31.

else is a recognized expert or if, in time of crisis, someone else has assumed the role of the leader.

The *withdrawal communicator* believes that nothing can be done to improve the situation at hand, and therefore the status quo should be maintained. New information can never benefit the situation but can make it worse. The withdrawal communicator generally stresses the past state of the art rather than the current or future state of the art.

We can now summarize the four philosophies:

- Controlling philosophy: "I wish to exert the greatest influence. I will tell you what to do and you will do it. Do not question my instructions.
- Developmental philosophy: "Let's work together to come up with the best approach. You try to convince me that your idea is correct and I'll try to convince you that my idea is best."
- Relinquishing philosophy: "You tell me what to do and I'll do it. I respect your authority and expertise."
- Withdrawal philosophy: "I'm very happy with the way we've been doing things. I'm not going to try to get you to do things differently and I don't want you to try to change my way of thinking. I wish to be uninvolved."

SUMMARY

In this chapter we discussed the nature and complexities of communication and presented the ideas behind communication networks, models, barriers, and techniques. Then we dealt with the subjects of informal communication and the role of executives, subordinates, customers, and culture in communication. We closed the chapter with a discussion of how to conduct meetings, write reports and memoranda, deliver a speech, conduct an interview, and develop a philosophy on communications.

In Chapter 3 we will discuss leadership in team management.

BIBLIOGRAPHY

Bavelas, Alex, and Dermot Barrett. "An Experimental Approach to Organizational Communication." *Personnel*, March 1951.

Buchwald, Art. *Reign of Terror: The Secretary's the Dictator*

Carlisle, Howard M. *Management: Concepts and Situations*. SRA Publishers, 1976.

Certo, Samuel C. *Principles of Modern Management*. Dubuque, Iowa: Brown, 1980.

Davis, Keith. *Human Behavior at Work*. New York: McGraw-Hill, 1972.

Gilchrist, J. C., M. E. Shaw, and L. C. Walker. "Some Effects of Unequal Distribution of Information in a Wheel Group Structure." *Journal of Abnormal and Social Psychology*, **49** (1954).

Harris, Philip R., and Robert T. Moran. *Managing Cultural Differences*. Houston: Gulf Publishing, 1979.

Jay, Anthony. "How to Run a Meeting." *Harvard Business Review*, March–April 1976.

Kerzner, Harold. *Project Management: A Systems Approach to Planning, Scheduling and Controlling*. New York: Van Nostrand Reinhold, 1979.

"Knowing How to Listen Takes More Than Two Good Ears." *Wall Street Journal*, September 15, 1980.

Leavitt, Harold J., and Ronald A. H. Mueller. "Some Effects of Feedback on Communicating." *Human Relations*, **4**(4) (November 1951).

McKenzie, Cheryl L., and Carol J. Qazi. "Communication Barriers in the Workplace." *Business Horizons*, March–April 1983.

Naisbitt, John. *Megatrends*. New York: Warner Books, 1982.

Nichols, Ralph G. "Listening Is Good Business." *Management of Personnel Quarterly*, Winter 1962.

Nichols, Ralph G. and Leonard A. Stevens. "Listening to People." *Harvard Business Review*, September–October 1957.

Peters, Thomas J., and Robert H. Waterman, Jr. *In Search of Excellence*. New York: Harper & Row, 1982.

Shaw, Malcolm E. *Developing Communication Skills*. West Point, Conn.: Educational Systems and Designs, Inc., 1968.

Stoner, James A. *Management*. Englewood Cliffs, N.J.: Prentice-Hall, 1982.

"Ten Commandments of Good Communications." *Management Review*, October 1955.

Walton, Eugene. "Communication Down the Line: How They Really Get the Word." *Personnel*, July–August 1959.

3
LEADERSHIP

HISTORICAL VIEW OF LEADERSHIP

The concept of leadership is an ancient one. All societies across the centuries, including the most primitive and the most civilized, have identified leaders and have developed one or more processes through which leaders emerge. Even in relatively simple societies leaders emerge in different areas of group life. Most primitive societies have at least two types of leaders, the political and the religious.

As long ago as the golden age of Greece, Plato theorized about the three types of leaders necessary for society:

1. The philosopher-statesman to rule the republic with reason and justice
2. The military commander to defend the state and enforce its will
3. The businessman to provide for the material needs of the citizens

Throughout history, "great men and women" representing Plato's three types have been celebrated and studied. Eras have been dominated by one type or another, leading historians to theorize that while greatness lies within the character of the individual, the culture provides the socializing experiences and the environment within which the "great person" can emerge.

The processes through which leaders can emerge differ from society to society. In many less complex societies, leadership is inherited and the leader makes decisions in many areas of community life. In more complex societies many types of leaders can emerge in different areas of community life. The role of institutions such as the Harvard Business School on the development of business leaders has been documented, and family and social ties have been shown to be a significant influence in the career of many leaders.[1]

[1] E. E. Jennings, *An Anatomy of Leadership: Princes, Herdes, and Supermen* (New York: Harper; 1960).

In most societies today, great interest is shown in community leaders, but no society has shown as much interest in the concept of leadership as our own. Beginning early in this century with the American sociologist, Cooler, who studied leaders of social movements, and continuing until the present with such studies as those presented by Peters and Waterman in *In Search of Excellence,* Americans have developed a substantial body of theory and practical information about leadership.[2,3]

Much of the recent research has been directed toward the identification of leadership skills and the development of training programs to teach these skills. The implications of much of the "new" thinking about leadership are that, as a society, we are moving rapidly to the necessity of a "team leadership" approach to many facets of community life. The need to bring to the decision-making structures of society people representing a variety of skills, knowledge, and background has resulted in widespread interest in the development of leadership skills in blacks, Hispanics, women, and others who have previously been excluded from these structures. Similarly, in business and industry the need for a wide representation of technical knowledge and skills on leadership teams has created a necessity to develop leadership skills as part of the technical skills armamentarium.

DEFINITIONS OF LEADERSHIP

In contemporary American social science literature there are four classes of definitions of leadership. The first, from an historical perspective, is leadership as a focus of group process. This class of definition stems from the study of social movements early in this century. Cooley postulated that "the leader is always the nucleus of a tendency and all social movements will be found to consist of tendencies having such nuclei."[4] This concept of leadership remains an important one today and can be seen in both large (union activities) and small (office party) social movements. Certain people seem to emerge spontaneously within the context of social activities and there seems to be an interaction between the value system of the individual, the priority he assigns to the activity, and the environmental context.

The second class of definitions may be grouped under the general heading of personal social control, a concept developed by Gordon Allport.[5]

[2] C. H. Cooler, *Human Nature and the Social Order* (New York: Scribners, 1902).
[3] T. J. Peters and R. H. Waterman, Jr., *In Search of Excellence* (New York: Harper & Row, 1982), pp. 137–323.
[4] C. H. Cooley, *Human Nature and the Social Order* (New York: Scribners, 1902).
[5] G. Allport, *Social Psychology* (Boston: Houghton Mifflin, 1924).

Personal social control is the individual leader's exercise of influence of his or her ability to motivate, persuade, or otherwise encourage followers toward a mutually agreed upon goal. A great deal of research on personal social control has been done, especially by psychologists and political scientists. The fascination with charismatic leaders in both popular and learned journals is a reflection of the interest in leadership as personal social control.

The concept of power is an integral aspect of personal social control. The sources of personal power have been identified by such authors as Benne, French and Raven, Blau, and Bennis.[6,7,8,9] These sources range from legitimacy—the acknowledged right to exercise power—to the authority of competence, to the ability to penalize and/or reward. Much of the study of bureaucracy has focused on the structure of power relationships. Many researchers have concluded that the overreliance on formal power relationships as a by-product of bureaucracy negates the potential benefits of such structures, and personal power derived from bureaucracy may be extremely unstable.[10] Personal power in community power structures appears to derive from a network of relationships of influence and favor granting based in the business and political arenas of community.[11,12] These findings would seem to indicate that personal social power implies a relationship of interdependency, influence, and exchange. In the absence of an environment of mutual obligation, there is no stable basis for the exercise of personal power. These concepts currently tie together the theories of personal social control with those of interaction and exchange, although initially the focus was on the individuals and secondarily on the sources of power rather than the content of the power relationship.[13,14,15,16]

In both of these definitions the analytic eye of the researcher was focused for the most part on the individual who had emerged or been ap-

---

[6] K. D. Benne, *A Conception of Authority* (New York: Teachers College, Columbia University, 1943).

[7] J. R. P. French and B. Raven, *The Bases of Social Power* (Ann Arbor: Studies in Social Power, University of Michigan Institute for Social Research, 1959).

[8] P. Blau, *Exchange and Power in Social Life* (New York: Wiley, 1964).

[9] W. Bennis, *American Bureaucracy* (Chicago: Aldine Press, 1970).

[10] R. Merton, *Social Theory and Social Structure* (New York: Free Press, 1967).

[11] F. Hunter, *Community Power Structure* (Chapel Hill: University of North Carolina Press, 1953).

[12] R. A. Dahl, *Who Governs?* (New Haven: Yale University Press, 1961).

[13] L. L. Bernard, *An Introduction to Social Psychology* (New York: Holt, 1926).

[14] W. Bennis, *American Bureaucracy* (Chicago: Aldine Press, 1970).

[15] J. W. Julian, *Leader and Group Behavior as Correlates of Adjustment and Performance in Negotiation Groups,* Dissertation Abstract, 1965.

[16] E. E. Jennings, *An Anatomy of Leadership: Princes, Herdes, and Supermen* (New York: Harper 1960).

pointed as a leader. The personality traits of "successful" leaders were and are studied extensively.[16] This preoccupation continues as we develop new social scientists such as psychohistorians and urban anthropologists. These studies of personalities and personal traits have extended to families that have produced many leaders over several generations. It has been speculated that leader-producing families are the American equivalent of the British aristocracy. Families such as the Rockefellers and Kennedys in our times have produced a disproportionate number of leaders in the political, business, and social reform arenas.

The third class of definitions of leadership relate to leadership "behaviors" rather than to individuals or specific social processes. Fiedler defined leadership as "leadership" behaviors by which we mean the particular acts in which a leader engages in the course of directing and coordinating the work of his group members.[17] Fiedler's contingency theory of leadership defines the functionality of leadership behaviors in terms of the situations and opens the way for the view that leadership may, in fact, be a set of behaviors all of which may be learned. Accordingly, if leadership behaviors can be identified as a group of skills or as groups of skills, and each group can be shown to be functional in a particular situation, people can be trained to be effective leaders.

The fourth class of definitions of leadership encompasses both the leaders and the followers by focusing on the interactions, expectations, and exchanges between leaders and followers. Blau defined leadership as the acknowledged superiority of the abilities of one member of the group which benefits the entire group.[18] The whole group benefits through both their leader's abilities and their support of him or her. Argyris postulated that "an organization will be most effective when its leadership provides the means whereby followers may make a creative contribution to it as a natural outgrowth of their needs for growth and self-expression."[19]

These latter definitions are especially helpful to people interested in organizations and management as they tend to focus on leadership skills in practical, everyday situations. Many of the researchers and theorists are business and organizational behavior scholars. Leadership in work groups tend to be the unit of analysis rather than leadership of more total social movements, professions, or business and political entities. The importance of leadership skills in everyday managerial functions was developed conceptually be Etzioni in *Modern Organizations* and refined by

---

[17] F. E. Fiedler, *A Contingency Model of Leadership Effectiveness* (New York: Academic Press, 1964).
[18] P. Blau, *Exchange and Power in Social Life* (New York: Wiley, 1964).
[19] C. Argyris, *Interpersonal Competence and Organizational Effectiveness* (Homewood, Ill.: Irwin-Dorsey, 1962), p. 81.

him in *Complex Organizations*.[20,21] The role of the manager in interfacing in the most productive manner the individual and the organization has become a legitimate area of leadership research.

These varying perspectives on leadership enable us to understand the complex nature of leadership and the diversity of traits and behaviors characteristic of different types of levels of leaders.

TYPES AND FUNCTIONS OF LEADERS

Leadership typologies have characterized research by sociologists and political scientists. More recently business scholars have begun to examine the effects of leadership types on certain work functions.

Traditionally leadership research classified leaders along such dimensions as autocratic-democratic, charismatic-bureaucratic, and task-instrumental. Leadership types have been developed by many scholars and classified by Stogdill as usually falling into the following categories:[22]

- Authoritative (dominator)
- Persuasive (crowd arouser)
- Democratic (group developer)
- Intellectual (eminent person)
- Executive (administrator)
- Representative (spokesman)

Business scholars have tended to examine the functional effect of leadership. Beginning with such works as R. C. Davis's *Fundamentals of Top Management,* business scholars have focused on the relationship between leadership and management in such functions as planning, organizing, and controlling.[23] Gross expanded the functions of business leaders to[24]

- Define goals
- Clarify goals
- Choose appropriate means
- Assign and coordinate facts
- Motivate

[20] A. Etzioni, *Modern Organizations* (New York: Free Press, 1964).

[21] A. Etzioni, *Complex Organizations* (New York: Free Press, 1972).

[22] R. M. Stodgill, "Personal Factors Associated with Leadership: A Survey of the Literature," *Journal of Psychology,* 1948, pp. 25, 33–71.

[23] R. C. Davis, *The Fundamentals of Top Management* (New York: Harper, 1951).

[24] E. Gross, "Dimensions of Leadership," *Personal Journal,* **40** (1961):213–218.

- Create loyalty
- Represent the group
- Spark action

Research which combines leader types with leader functions has proven most productive. Indications are that more democratic approaches are effective where technical skills are dispersed (an R&D division in industry), whereas more authoritative approaches are effective in situations where technical skills are hierarchical (an emergency room in a hospital.[25] Chris Argyris conceptualizes the leadership role in management as one that mediates the needs of the organization with the needs of the employees and is modified accordingly.[26] The presumption is that when the employees are motivated to meet their own needs within the context of the needs of the organization, the rewards will be maximized for both.

As part of the history of the development of management as leadership research, the role of the National Training Laboratories must be mentioned. Led by a group of businessmen greatly influenced by humanistic psychology, the National Training Labs began to offer "sensitivity" or "empathy" training to business executives based on the notion that opening communication and developing empathic skills would enhance the capacity to manage. This confusion between managing and leading and between open communication and leadership skills has persisted for some time and is still in common use despite research which indicated there is usually no relationship and sometimes an inverse relationship.[27] The vogue of T (for training) groups is indicative of the sometimes unfortunate manner in which serious subjects are commercialized into triviality. The other side of that problem is the loss of substantial, serious research because of the inability of the scholars to reach the appropriate market. In this as in other areas of science and technology, the transfer of knowledge is an important issue.

Training in both the art and technical skills of leadership should be integrated into the education of all professionals as well as those who are specializing in management. Since leadership implies relationships with other people, these "people" skills need to be raised in prestige to the same level as other technical and decision-making skills.

---

[25] F. E. Fiedler, *A Contingency Model of Leadership Effectiveness* (New York: Academic Press, 1964).

[26] C. Argyris, *Interpersonal Competence and Organizational Effectiveness* (Homewood, Ill.: Irwin-Dorsey, 1962), p. 81.

[27] K. Back, *Beyond Words* (New York: Wiley, 1969).

LEADERSHIP TRAITS

At the beginning of this century, leaders were generally regarded as superior individuals. They possessed qualities and abilities that differentiated them from the general populace. These qualities and abilities were attributed to inheritance or social adventure. The search for the specific qualities occupied the succeeding generations.

The qualities found to be associated with leadership are as follows:

- Capacity (intelligence, alertness, verbal facility, originality, judgment)
- Achievement (scholarship, knowledge)
- Responsibility (dependability, initiative, persistence, aggressiveness, self-confidence, desire to excel)
- Participation (activity, sociability, cooperation, adaptability, humor)
- Status (socioeconomic position, popularity)
- Situation (mental level, status, skills, needs and interests of followers, objectives to be achieved, etc.)

Stogdill cited evidence that leadership is entirely situational in origin and that no personal characteristics (age, height, weight, physique, energy, appearance, etc.) will predict leadership.[28]

Leadership is a position of responsibility for coordinating the activities of the members of the group in their task of attaining a common goal. There is a preponderance of evidence which indicates that leadership traits differ with the situation. The qualities, characteristics, and skills required in a leader are determined by the demands of the situation in which he or she is to perform as a leader. A person, therefore, does not become a leader by virtue of some combination of the above six qualities alone, because the characteristics of the leader must bear some relevant relationships to the characteristics, activities, and goals of the followers.

Leadership is the interaction of many variables which are in constant flux and change set by the demands of the situation—addition or loss of members, changes in interpersonal relationships, changes in goals, competition of outside influences, etc. The very studies which provide the strongest arguments for the situational aspect of leadership also supply the strongest evidence indicating that leadership qualities are also persistent and relatively stable.[29]

---

[28] R. M. Stogdill, "Personal Factors Associated with Leadership: A Survey of the Literature," *Journal of Psychology,* 1948, pp. 25, 33–71.

[29] W. I. Newstetter, M. J. Feldstein, and T. M. Newcomb, *Group Adjustment: A Study in Experimental Sociology* (Cleveland: Western Reserve University, 1938).

BEHAVIOR PATTERN

The behavior pattern of a leader must be considered because, as stated above, (1) little success has been attained in attempts to quantify leaders in terms of traits, (2) traits demanded in a leader varied from one situation to another, and (3) the trait approach ignores the interaction between leader and followers.

Halpin and Winer measured two distinct patterns of behavior as key to describing leadership:[30]

- Consideration—This is the extent to which a leader exhibits concern for the welfare to other group members. Considerate supervisors exhibit appreciation of good work, stress the importance of job satisfaction, maintain and strengthen self-esteem of subordinates by treating them as equals, make special efforts for subordinates to feel at ease, are easy to approach, put subordinates' suggestions into operation, and obtain concurrence on important matters.
- Initiation of structure—This is the extent to which a leader initiated activity in the group, organized it, and defined the subsequent course of operation. Included in the initiation of structure is the behavior of maintaining standards and meeting deadlines, deciding in detail on the course of action and how it should be accomplished. The leaders' own role, and those of subordinates, toward goal attainment is particularly important.

Consideration and initiation of structure are the leader behaviors that have received the greatest attention from behavioral scientists. However, the following behaviors may form important additions to a leader's repertoire:

- Personally rewarding
- Personally punishing
- Goal setting
- Designing feedback systems
- Placing personnel
- Designing job systems

There is considerable evidence that leader-reward behavior has an important impact on subordinates. Leaders who identify good performance

[30] A. W. Halpin and B. J. Winer, *A Factorial Study of the Leader Behavior Description* (Columbus: Ohio State University, Bureau of Business Research, 1957).

and subsequent rewards (compliments, tangible benefits, or special treatment) tend to enhance subordinate satisfaction and performance. With such leadership, subordinates have a clear picture of what is expected of them and anticipate positive outcomes if they achieve these expectations.

Personally punishing behavior involves the use of reprimands and the

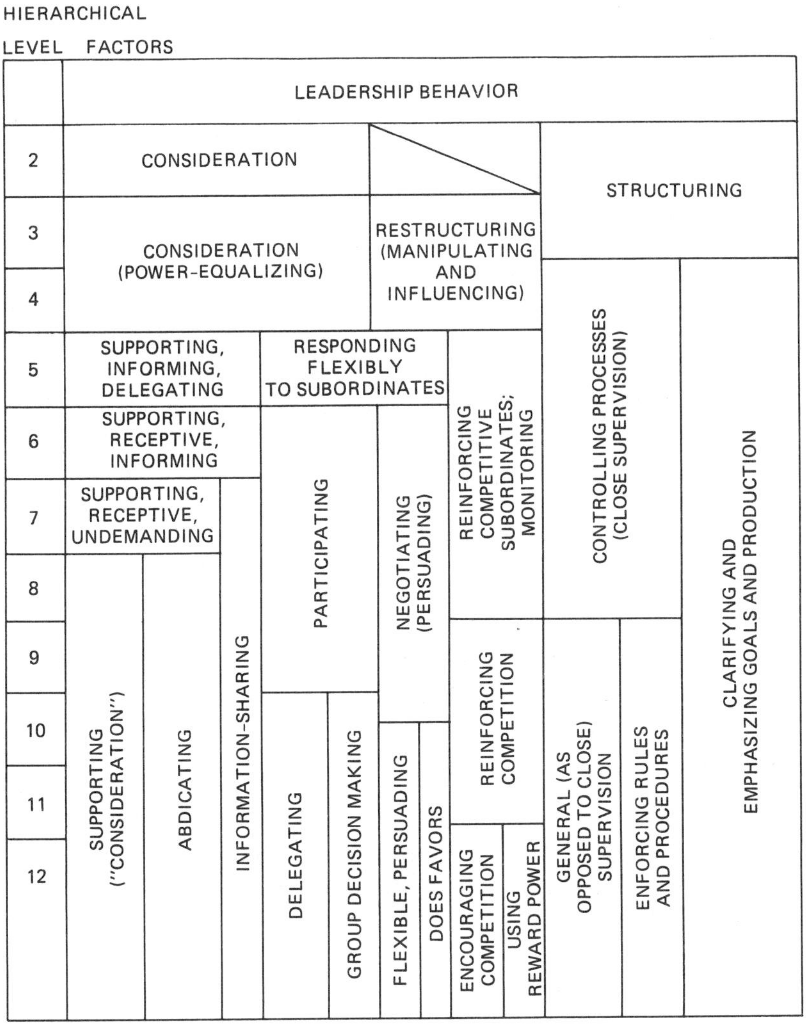

HIERARCHICAL LEVEL | FACTORS

LEADERSHIP BEHAVIOR

2 CONSIDERATION | STRUCTURING

3–4 CONSIDERATION (POWER-EQUALIZING) | RESTRUCTURING (MANIPULATING AND INFLUENCING)

5 SUPPORTING, INFORMING, DELEGATING | RESPONDING FLEXIBLY TO SUBORDINATES

6 SUPPORTING, RECEPTIVE, INFORMING

7 SUPPORTING, RECEPTIVE, UNDEMANDING

8–12: SUPPORTING ("CONSIDERATION") | ABDICATING | INFORMATION-SHARING | PARTICIPATING | DELEGATING | GROUP DECISION MAKING | NEGOTIATING (PERSUADING) | FLEXIBLE, PERSUADING | DOES FAVORS | REINFORCING COMPETITIVE SUBORDINATES; MONITORING | REINFORCING COMPETITION | ENCOURAGING COMPETITION | USING REWARD POWER | CONTROLLING PROCESSES (CLOSE SUPERVISION) | GENERAL (AS OPPOSED TO CLOSE) SUPERVISION | ENFORCING RULES AND PROCEDURES | CLARIFYING AND EMPHASIZING GOALS AND PRODUCTION

Figure 3-1. The hierarchical structure of leadership behaviors. (From J. A. Miller, *A Hierarchical Structure of Leadership Behavior,* Rochester, N.Y.: University of Rochester, Management Research Center, Technical Report No. 66, 1973.)

active withholding of raises or promotions when subordinate performance is inadequate. In some cases, punishment appears to contribute to subordinate dissatisfaction and ineffectiveness. But in others, it is associated with favorable subordinate response depending on the leader's skill in using punishment effectively.

There has been little systematic investigation of the remaining leader behaviors (setting goals, designing feedback systems, placing personnel, and designing job systems). These, however, should be effective in leader behavior.

Miller developed a hierarchical solution to leadership behavior, as shown in Figure 3-1, which encompasses some of the above considerations.[31]

LEADERSHIP STYLES

Leaders (managers or administrators) do more than just supervise subordinates. The leaders' effectiveness depends to a considerable degree on getting work done through others. Interpersonal skills are needed at all levels. A leader's hierarchical level, Figure 3-1, affects the time for which decisions are to be made and the feedback lag about the effects.

Mahoney, Jerdee, and Carroll recognized eight important functions of a manager—planning, investigating, coordinating, evaluating, supervising, staffing, negotiating, and representing.[32] Supervising along with the four functions of planning, investigating, coordinating, and evaluating, account for almost 90% of all the time spent at work by most managers.

The single most important function of a first-line manager is supervising others, including dealing with disciplinary problems and problems of promoting, rating, and classifying employees.

Mackenzie illustrated the great variety of activities that may be performed by a typical manager.[33] Figure 3-2, shows the elements, tasks, functions, and activities that may be part of a manager's job. At the center or hub are people, ideas, and things—the basic components of every organization. According to Mackenzie, ideas create the need for conceptual thinking; things, for administration; people, for leadership.

Problem analysis, decision making, and communications permeate all aspects at all times of the jobs held by managers. Other functions usually

---

[31] J. A. Miller, *A Hierarchical Structure of Leadership Behavior,* Rochester, N.Y.: University of Rochester, Management Research Center, Technical Report No. 66, 1973.

[32] T. A. Mahoney, T. H. Jerdee, and S. T. Carroll, "The Job(s) of Management," *Industrial Relations,* 1965, pp. 97–110.

[33] R. A. MacKenzie, "The Management Process in 3-D," *Harvard Business Review,* **47** (1969):80–87.

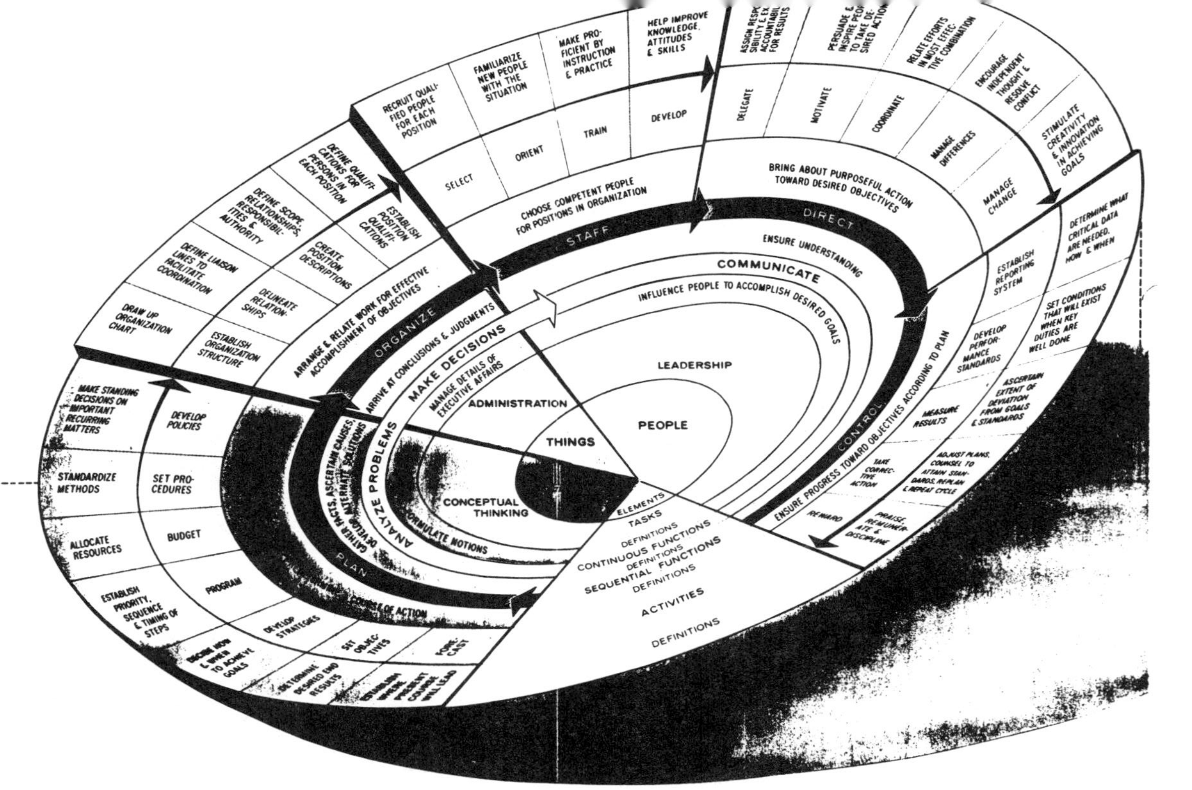

Figure 3-2. Leadership and management elements, tasks, functions, and activities. *Source: Harvard Business Review*, "The Management Process in 3-D," by R. Alec MacKenzie (November-December 1969). Copyright © 1969 by the President and Fellows of Harvard College; all rights reserved.)

occur in a more predictable sequence; thus, planning, organizing, staffing, directing, and controlling are shown, in that order, in one of the bands. Position and the stage of completion of the project dictate the involvement of managers in these sequential functions. The relevant activities and their definitions are shown on the outermost bands of the diagram. Managers must sense and be in tune with the pulse of their organization.

The diagram clearly shows the sheer diversity of activities and their linkages to functions and tasks of management (leadership). It also points up the inadequacies of any simple approach in trying to formulate what is involved in the managerial process. The motivational needs of the subordinates must also be recognized by the leaders. This will be covered in Chapter 4.

PARTICIPATIVE LEADERSHIP

One of the most persistent and controversial issues in the study of leadership concerns employee participation in the decision-making process. Traditional models of leadership are autocratic in nature.

Behavioral scientists who have studied the issue have concluded that participation by subordinates in the problem-solving and decision-making process is positive. Pointing to evidence of restriction of output and lack of involvement under traditional (directive) leadership styles, they argue for greater influence in decision making on the part of those who are held responsible for decision execution.

Participation has two distinct meanings, one encompassing the other:

1. The leader-subordinate decision making in which there is power equalization by the leader and a sharing of final decision. Consensus is sought.
2. The general expression to refer to the half of the continuum of decision making in which subordinates are involved in some way in the decision process, either because they are consulted by the leader (who makes the final decision), or they share in making the final decision, or they are delegated decision responsibility by the leader.

Vroom and Yetton have written extensively on this subject, and they indicate that empirical evidence provides some, but not overwhelming, support for beliefs in the efficacy of participative leadership.[34] Field experiments indicate that increases in productivity can be brought about by

[34] V. H. Vroom and P. W. Yetton, *Leadership and Decision-Making* (Pittsburgh: University of Pittsburgh Press, 1973).

giving subordinates an opportunity to participate in decision making and goal setting.

The degree of participation for effective leadership methods or style depends on the situation and to some extent on personality. Advocates such as Argyris have noted that "situational relativity" of leadership styles. Thus he writes:

> No one leadership style is the most effective. Each is probably effective under a given set of conditions. Consequently, I suggest that effective leaders are those who are capable of behaving in many different leadership styles, depending on the requirements of reality as they and others perceive it. I call this "reality-centered" leadership.[35]

It is recognized that different situations require different leadership methods. There is, however, less agreement concerning what is appropriate for the analysis of the situation.

Participative leadership requires a leader with power who is willing to share it. With his or her power, the leader sets the boundaries within which a subordinate's participation or consultation is welcomed.

The key leadership challenge for the 1980s is the adaptation of participative leadership with recognition of the situational aspects and ability to adapt. Situational leadership can apply to both formal and informal leaders as shown in Figure 3-3. In Part B of Figure 3-3, the informal leader may be the project manager, whereas the formal leader is the line manager.

ELEMENTS OF CULTURE

Culture is involved in many aspects of human life. Within the turbulent business world, culture can be considered the ideas and standards that a group of people have in common. Two aspects of culture seem to be particularly applicable to the business community: environment and values. The influence of environment within the cultural context has been studied in some depth by Lawrence and Lorsch.[36] Some of the better understood factors of environment are

- Type of business
- Market

---

[35] C. Argyris, *Interpersonal Competence and Organizational Effectiveness* (Homewood, Ill.: Irwin-Dorsey, 1962), p. 81.

[36] P. Lawrence and J. Lorsch, *Organizations and Environment* (Cambridge: Harvard University Press, 1967).

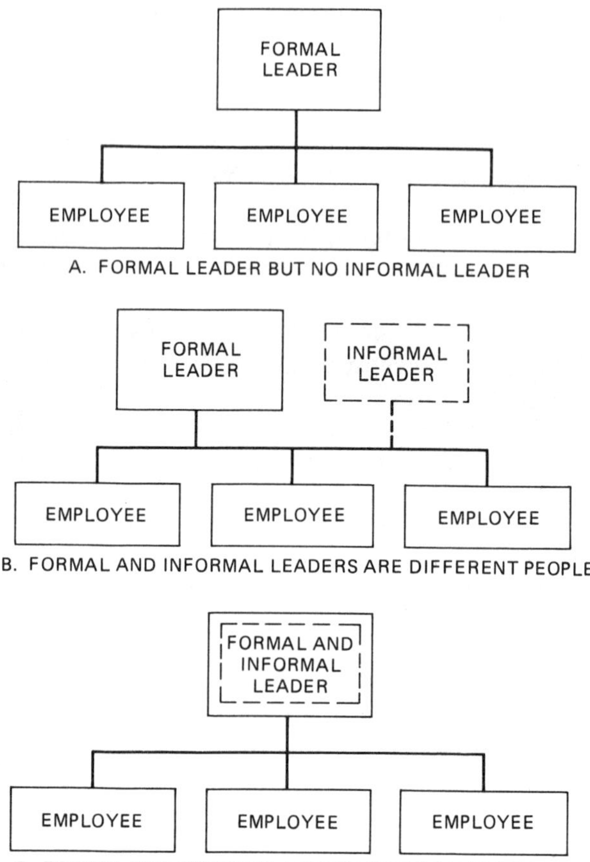

Figure 3-3. Formal and informal leaders.

- Competition
- Government
- Region

These factors as well as many others typically must be taken into consideration for decisions at the upper management levels. Each of the elements above can assume a wide variety of values, combinations of which can produce an enormous number of possible situations and probable solutions. One of the advantages of a culture is its effect on the thinking process. For a given situation there is a limited number of alternatives which seem plausible even in a surface examination. This tends to reduce the amount of resources that will be expended in order to make a choice among possible alternatives.

The second major category of culture is that of values. Values originate from the environment, usually from families or society. Kumar has done a study of values of college students.[37] That study seemed to indicate that the perception or needs of wants was largely influenced by values originating from the student's background. Culture can be viewed as a set of values to aid in determining what is right and wrong or what needs to be emphasized, approved, or disapproved. It is pointed out by Dowling that the relative importance of certain needs or wants is often a result of the society in which a person lives.[38] Since environment and values are so closely interrelated, they often dictate what makes sense for a specific organization in a given economic environment.

In the business environment, values provide a sense of direction for guiding a large number of daily activities. Because values are so heavily dependent on the situation, it is often difficult to make an objective study or comparison of values unless the situations are very similar. For organizations to have a high probability of success, they must make a deliberate effort to ensure that at least some of the organization's values are acceptable or compatible with the values of the people within the organization. Values can also have a significant effect on careers. If a company's values are particularly strong in one functional area, that area will probably see a larger-than-average share of promotions. There is, however, some danger in letting a small number of values dominate a company's culture. If one of those values falls out of favor with the market, the company can experience severe trauma in trying to adapt to the changing environment. Another problem with trying to promote a set of values is that it is very difficult to avoid unintentional contradictions as information travels from one group of people to another.

THE MODERN MANAGER

A great deal of research was conducted about what effective business leaders do, and the findings were very different from the traditional managerial approach. The following, based on research conducted by H. Mintzberg, are a reflection of what current business managers do.[39]

- Most managers are not reflective planners but are continually busy with interpersonal contacts.

---

[37] P. Kumar, *Psychology Studies,* 1965, 10(2), 73–79.
[38] J. Dowling and J. Pfeffer, *Pacific Sociological Review*. 1975, 1, 122–136.
[39] H. Mintzberg, *The Nature of Managerial* (New York: Harper & Row, 1973).

- Much of the manager's time is spent not in decision making, but in obtaining information.
- A large part of a manager's performance is controlled by other individuals and groups.
- Most information gathered is by word of mouth.
- Skills required of a manager are significantly different from those taught in business schools.
- Planning amid chaos and making judgments on oral information appear to be significant.
- The manager needs to be able to negotiate with people outside the organization for necessary resources.
- Managers do not plan or organize to the extent traditional management theorists taught.

Recent teachings about leadership are probably best expressed by E. P. Hollander, who emphasized that leadership is a two-way influence pro-

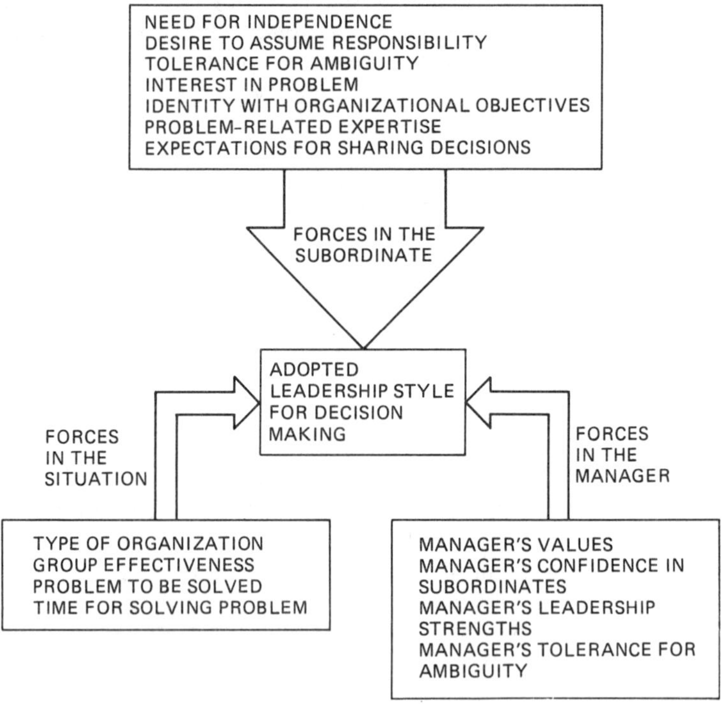

Figure 3-4. Collective influence of forces in the manager, subordinates, and the situation on the leadership style adopted for decision making.

cess involving a social exchange relationship between the leader and his followers.[40] The relationship involves the leader providing direction, acting as a buffer in conflict resolution, and helping to reduce uncertainty because of ambiguity. In exchange, the leader obtains the efforts of the followers. The transactional leadership approach is based on the assumption that all instances of successful leadership are different and require unique combinations of leaders, followers, and leadership situations. (See Figure 3-4.) This theory of interaction is often expressed by the formula SL = f(L,F,S).

> In this formula, SL is successful leadership, f stands for function of and L, F, and S are, respectively, the leader, the follower, and the situation. A translation of this formula would be that successful leadership is the function of the leader, the follower, and the situation. In other words, the leader, the follower, and the situation must be appropriate for one another if a leadership attempt is to be successful.[41]

LEADERS IN THE 1980s

Leadership education in the 1980s centers around the interaction between the leader and the follower. Figure 3-5 deals graphically with the teachings of E. P. Hollander and demonstrates the need for equal and mutual benefit of the company, leader, and the follower.[42] In order for the transaction process to function effectively, the benefits must be perceived as equitable. The growth of the transactional leadership approach in the late 1970s and 1980s has been due to the need to motivate followers, through effective communications, in order to achieve productivity improvements necessary to meet competitive pressures.

The most effective businesses are those which can coordinate the needs of the customers, organization, and employees through effective communication. However, from a leadership standpoint, the higher the manager moves in the organization, the more removed he is from the feedback loop.

Additional communication barriers are the interpersonal-hostility (I-H) barrier and the parliamentary-methods (P-M) barrier. Both can hinder the flow of valuable and timely information. The I-H barrier is evident when "I won't give that guy the satisfaction of admitting he's got a good idea,

---

[40] E. Hollander, *Leadership Dynamics: A Practical Guide to Effective Relationships* (New York: Free Press, 1978).
[41] S. C. Certo, *Principles of Modern Management* (Dubuque, Iowa: Brown, 1983).
[42] E. Hollander, *Leadership Dynamics: A Practical Guide to Effective Relationships* (New York: Free Press, 1978).

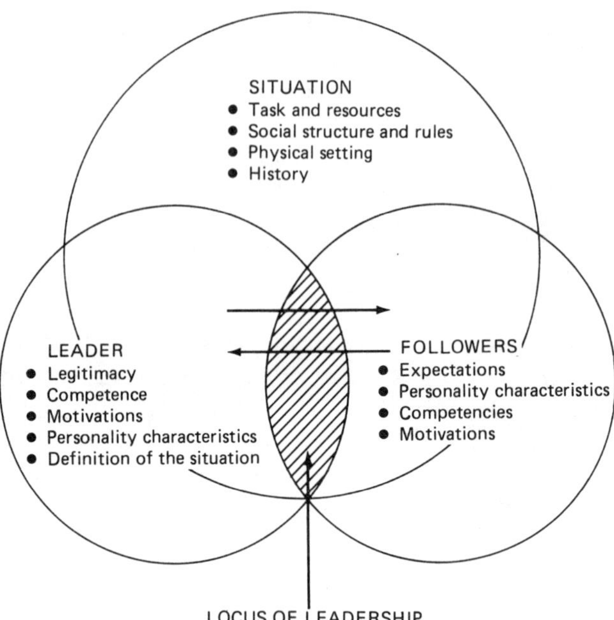

Figure 3-5. Three elements involved in leadership—the situation, the leader, and the follower—with some of their relevant attributes. The arrows indicate the social exchange which occurs there between the leader and the followers. (*Source:* Hollander, *Leadership Dynamics: A Practical Guide to Effective Relationships* (New York: Free Press, 1978), p. 8.)

and I certainly won't let him in on my idea." The P-M barrier is evident when "I can't speak until I'm recognized by the chair and if the chair never recognizes me, my information will never come out."

LEADERSHIP STUDIES

At Ohio State University, researchers studied the effectiveness of leadership behavior using the model shown in Figure 3-6. The "initiating structure" is equivalent to task-oriented behavior and "consideration" is employee-oriented behavior. The results of the study showed that the subordinates rated the effectiveness of their leader based more upon the situation than on the actual leadership style used. Furthermore, employee satisfaction was highest for leaders with stronger desires for employee-oriented behavioral management than for task-oriented behavior.

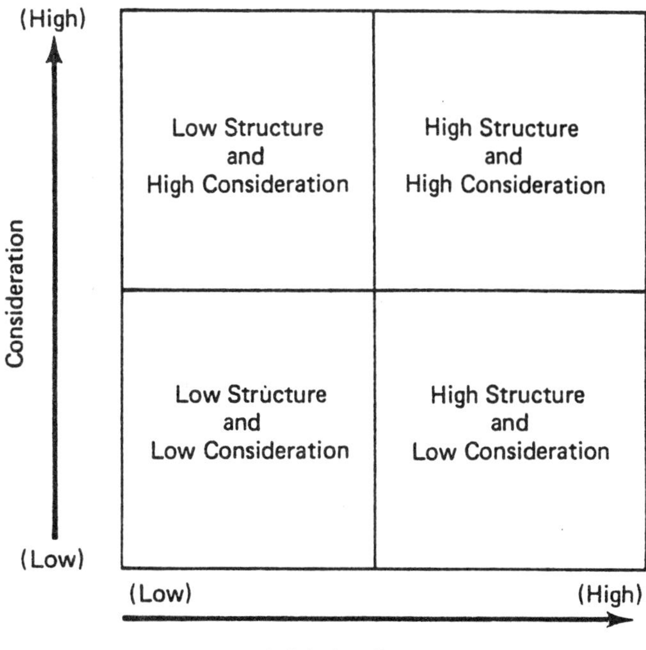

Figure 3-6. Leadership styles studied at Ohio State.

A similar study at the University of Michigan compared production-centered and employee-centered leadership styles. Production-centered managers stressed work standardization and rigid work flow requirements whereas employee-centered managers stressed group decision making and goal setting. As expected, subordinates preferred managers who were employee-centered because of the higher involvement by employees.

These two studies underlie the basis for the Managerial Grid developed by Blake and Mouton as shown in Figure 3-7.[43] The vertical axis is the leader's concerns for people, whereas the horizontal axis is the leader's concern for getting the job done. The level of concern on each axis is measured on a scale of 1 to 9. A leader's position on the grid is determined by analyzing various responses to a series of questions regarding his feelings toward the job and the people he must lead. Ideally, the 9, 9 corner is the ideal position for all managers to strive for.

[43] See R. R. Blake and J. S. Mouton, *The New Managerial Grid* (Houston: Gulf Publishing, 1978), p. 11.

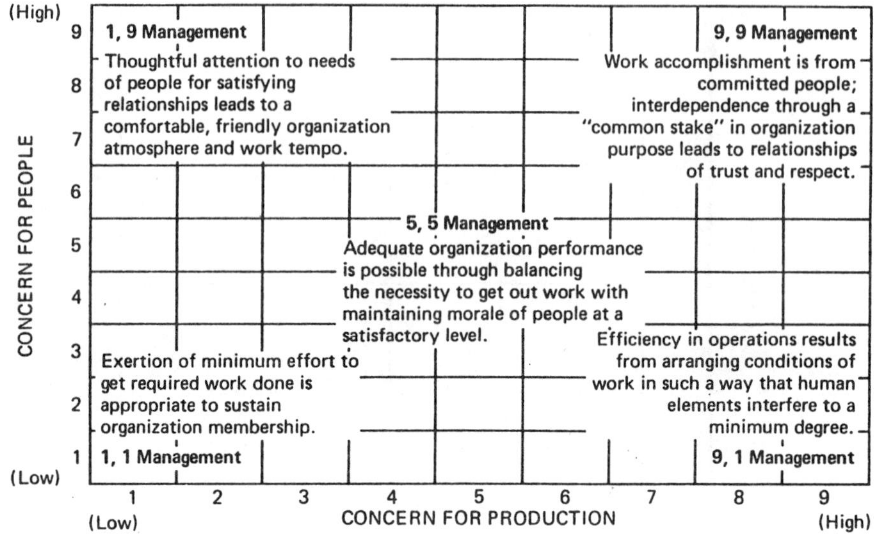

Figure 3-7. The managerial grid. (*Source: The New Managerial Grid* by Robert R. Blake and Jane Srygley Mouton (Houston: Gulf Publishing Company), p. 11. Copyright © 1978. Used by permission.)

SELECTING THE APPROPRIATE LEADERSHIP STYLE FOR PROJECT MANAGEMENT

Leadership is defined as a process whereby an individual attempts to accomplish desired goals by influencing others. Leadership is a subdivision of management where employees attempt to get work accomplished through others. Leadership involves an analysis of the leader, follower, and situation.

Leadership is extremely difficult in a project management mode (especially in project-driven companies) because (1) the project manager may not have the power to provide monetary and nonmonetary rewards, (2) the employees may not report directly to the project manager, (3) the employees may be sharing their time with other projects, (4) the project duration may be very short, and (5) the employees may never have worked for this project manager previously and may never work for him again.

The effective leader has multiple leadership styles depending upon the situation. The leader knows when to be tough and when to be soft, and must be willing to treat the same person differently if the situation is different.

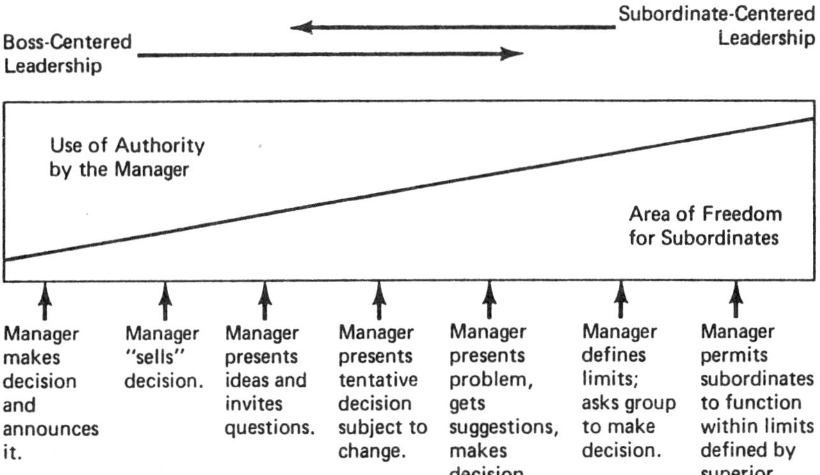

Figure 3-8. Continuum of leadership behavior. (*Source:* Robert Tannenbaum and Warren H. Schmidt, "How to Choose a Leadership Pattern," *Harvard Business Review,* 51, no. 3 (May–June 1973), p. 166. Copyright © 1973 by the President and Fellows of Harvard College; all rights reserved.)

The exact leadership style that project managers select is most often based upon their authority (See Figure 3-8).[44] Levels of authority have a strong bearing on whether or not a manager should be task-oriented or people-oriented. Project managers with a great deal of legal authority can be either task- or people-oriented, whereas project managers with very little authority tend to be more people-oriented.

The leader's background, education, experience, and values can have a dominant impact on the selection of a leadership style. Retired military officers who are placed in industrial project management positions may tend to adopt an authoritarian, task-oriented leadership style. Project managers who have strong ties with professional organizations may believe that their professional needs and ego trips are more important than the project or company.

The characteristics of subordinates must also be considered before selecting an appropriate leadership style. Tannenbaum and Schmidt have shown that the needs of the organizational personnel (i.e., subordinates), especially with respect to decision making, participation, and freedom,

[44] Robert Tannenbaum and Warren H. Schmidt, "How to Choose a Leadership Pattern," *Harvard Business Review,* **51**(3) (May–June 1973).

influence the boss-centered/subordinate-centered leadership role. Greater freedom can be given to subordinates when they[45]

- Crave independence and freedom of action
- Want to have decision-making responsibility
- Identify with the organization's goals
- Are knowledgeable and experienced enough to deal with the problem efficiently
- Have experience and previous working relationships with managers that leads them to expect participative management

This situation can also play an important role in selecting the appropriate leadership style. Although the Tannenbaum-Schmidt model of Figure 3-8 tends to imply that managers will have one leadership style in each situation, this may not hold true for project managers. With rapid changes in the environment, requirements for strong adherence to the time, cost, and performance constraints, and possible customer uncertainties, project managers may find it necessary to shift very rapidly from subordinate-centered to boss-centered leadership styles. This most frequently occurs in time of crisis.

Not all researchers believe that managers can effectively change their leadership style to fit the situation. Fiedler believes that leadership styles are inflexible and since no single leadership style is effective for all situations, effective performance can be achieved only by matching the manager to the situation or by changing the situation. According to Fiedler,

> A person who describes his least preferred co-workers (LPC) in a relatively favorable manner tends to be permissive, human relation-oriented and considerate in the feelings of his men. But a person who describes his least preferred co-workers in an unfavorable manner—who has what we come to call a low LPC rating—tends to be managing, task-controlling and less concerned with the human relations aspects of the job.[46]

Fiedler believes that the leadership effectiveness of managers can be directly related to their LPC scores as shown in Figure 3-9. High LPC managers desire close ties with subordinates, whereas low LPC managers are task-oriented and place a low priority on employee feelings.

---

[45] Ibid.

[46] Fred E. Fiedler, "Engineer the Job to Fit the Manager," *Harvard Business Review*, **43**(5) (September–October 1965):116.

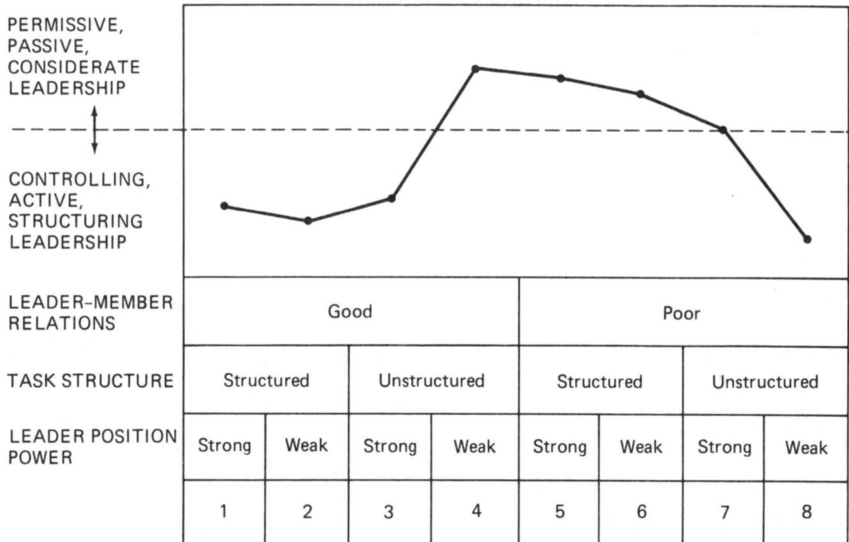

Figure 3-9. Fielder's contingency theory of leadership. (Copyright © 1984 by Houghton Mifflin Company. *Source:* Fred E. Fiedler, "The Effects of Leadership Training and Experience: A Contingency Model Interpretation," Administrative Science Quarterly, December 1972, p. 455. Used with permission.)

The Fiedler model identifies three situational work elements:

1. *Leader-member relations:* A manager who is trusted and respected by subordinates may not have to rely upon authority to get the job done.
2. *Task structure:* Managers with well-structured activities, where subordinates know what they are expected to do, generally have a great deal of authority and power. Unstructured activities provide nebulous guidance, a continuous stream of questions, insecurity, and reduced authority for the manager.
3. *Position power:* The location where an employee sits in the organizational hierarchy is regarded as position power. High position power usually simplifies the leader's ability to influence subordinates.

The Fiedler model suggests that these three situational work elements must be matched to the leader's management style as indicated by the LPC score. For example, if the leader is not trusted by the subordinates (i.e., poor leader-member relations), has a weak power base, and is managing an unstructured task, then it is best to select someone with a task-motivated leadership style (i.e., category 8).

The Fiedler model can be applied to project management. As described in Chapter 1 under desirable characteristics, project managers should have strong interpersonal skills, thus providing good leader-member relations. Project management advocates well-structured planning (using a statement of work, specifications, milestone schedules, and a work breakdown structure) thus providing task structure. Finally, project managers do not normally possess position power, but obtain a strong power base by being recognized as an expert in project management credibility with the employees and by being an effective decision maker. Therefore, ideal project managers would occupy category 1.

Most projects are well structured, and, with some degree of experience, project managers develop position power simply by being project managers. The weakness is in leader-member relations. According to the Fiedler model, therefore, category 1 or 5 is most likely. Category 1 is beneficial on long-term projects where the project manager can develop leader-member relationships. However, for many other projects, leader-member relationships may be poor because

- Functional employees spend only a small portion of their time on any one project.
- The time frame of the project is too short to develop meaningful relationships.
- Functional employees look toward their line managers for leadership rather than the project managers.

This last item is extremely important since, in a project environment, employees must report to one line manager and perhaps more than one project manager, all at the same time. Therefore, in a project management environment, the leadership of the subordinates may have to be shared between the project and line managers. If this is the case, then perhaps in a dual leadership mode, project managers should accept a leadership style that is compatible with the majority of *the line managers*.

LIFE-CYCLE LEADERSHIP

Perhaps the best model for analyzing leadership in a project management environment was developed by Hersey and Blanchard.[47] The model, as

[47] Paul Hersey and Kenneth Blanchard, *Management of Organizational Behavior* (Englewood Cliffs, N.J.: Prentice-Hall, 1979), p. 165.

shown in Figure 3-10, is the life-cycle theory of leadership. Hersey and Blanchard contend that leadership styles must change according to the maturity of the employees, with maturity defined as job-related experience, willingness to accept job responsibility, and desire to achieve. This definition of maturity is somewhat different from other behavioral management definitions which define maturity as age or emotional stability.

As shown in Figure 3-10, the subordinates enter the organization in the initial phase, which is high task and low relationships behavior. In the initial phase, the leadership style is almost pure task behavior and is an autocratic approach where the leader's main concern is the accomplishment of the objective, often with very little concern for the employees or their feelings. The leader is highly forceful and relies heavily upon his own abilities and judgment. Other people's opinions may be of no concern. Hersey and Blanchard assume that, in the initial stage, there is anxiety, tension, and confusion among new employees so that relationship behavior is inappropriate.

In the second phase, employees begin to understand their tasks and the leader tries to develop strong behavioral relationships. The development

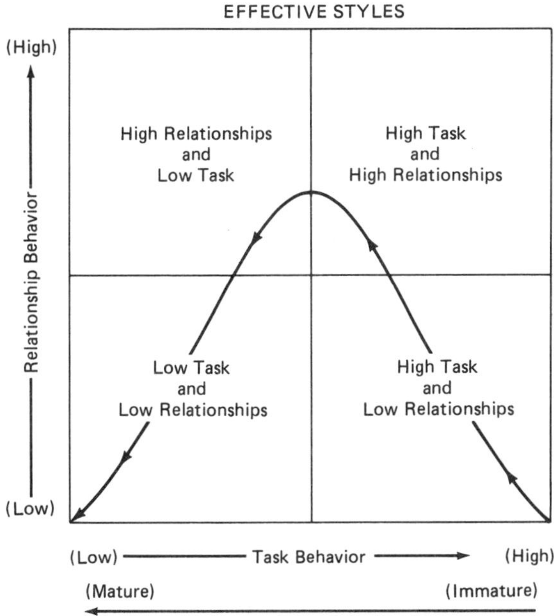

Figure 3-10. The life-cycle theory of leadership. (*Source:* Adapted from Paul Hersey and Kenneth H. Blanchard, *Management of Organizational Behavior,* 3rd ed., p. 165. Copyright © 1979. Used by permission of Prentice-Hall, Inc.)

of trust and understanding between the leader and subordinates becomes a driving force for the strong behavioral relationships. However, although the leader begins utilizing behavioral relationships, there still exists a strong need for high task behavior as well since the employees may not have achieved the level of competency to assume full responsibility.

The third phase is often regarded as pure relationship behavior where the leader is more interested in gaining the respect of the employees than perhaps in achieving the objectives. Referent power becomes extremely important. This behavior can be characterized by delegation of authority and responsibility (often excessive), participative management, and group decision making. Hersey and Blanchard believe that, in this phase, the employees no longer need directives and are knowledgeable enough about the job and self-motivated to the extent that they are willing to assume more responsibility for the task. Therefore, the leader can try to straighten his relationships with subordinates.

In the fourth phase, the employees are experienced in the job and confident about their own abilities, and are trusted to handle the work themselves. The leader demonstrates low task and low relationship behavior as the employees mature.

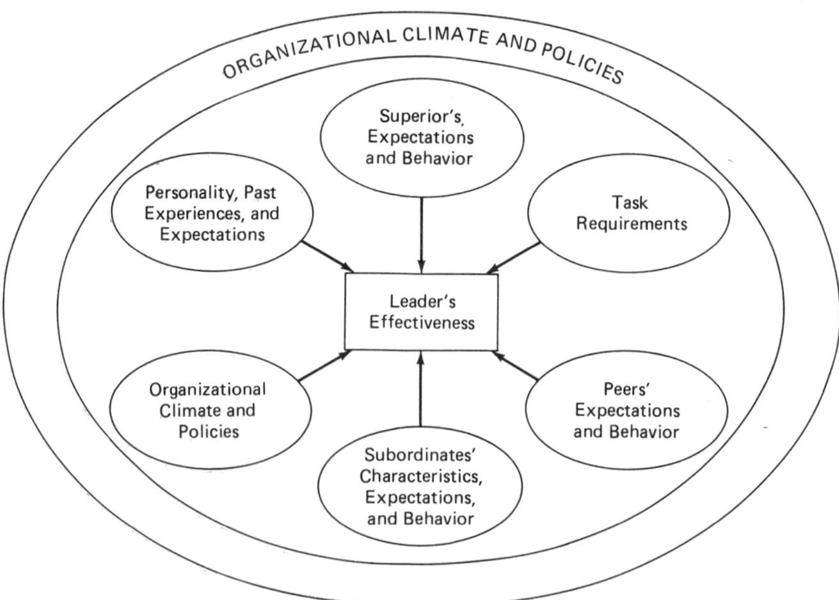

Figure 3.11. Personality and situational factors that influence effective leadership. *Source:* James A. F. Stoner, *Management*, 2nd Edition, Prentice-Hall Inc., Englewood Cliffs, N.J. Used by permission.

This type of life cycle approach to leadership is extremely important to project managers because it implies that effective leadership must be dynamic and flexible rather than static and rigid. (See Figure 3-11.) Effective leaders are neither pure task or relationship behavioralists, but maintain a balance between them. However, in time of crisis, the leader may be required to demonstrate a pure behavioral style or a pure task style.

LONG-TERM AND SHORT-TERM LEADERSHIP SKILLS

The manager of the 1980s must possess two sets of skills. The long-term skills are the functional skills patterned after traditional business school programs. These skills are used when deliberating or in the decision-making process. The second set of skills are short-term in nature and deal with attempting to direct the day-to-day work activities of employees. The development of communication skills to motivate followers is becoming increasingly important. Strong communication skills of the leader will result in an ability to (1) motivate and inspire, (2) stimulate cooperation and trust, (3) persuade, and (4) bargain and negotiate.

SITUATION MANAGEMENT SYSTEMS AND FORUM CORPORATION

Two educational programs cited here are available to assist in the development of leaders who will be able to increase the productivity of their followers. Situation Management Systems, Inc., provides a program that evaluates the leader's management style in the following areas:

- Use of reward and punishment techniques
- Use of participation and trust techniques
- Use of common vision techniques
- Use of assertive persuasion techniques

Selected followers are asked to provide data to an independent source about the various leadership characteristics of the manager. Several questions are asked in each category and the data compiled by SMS Inc. The data are then summarized and provided to the leader. The followers, due to the secrecy of the feedback, provide accurate leadership analysis of the manager. The data demonstrate what the needs of the followers are while advising the manager of his strong leadership skills and areas needing correction.

A second educational program is one offered by the Forum Corporation and stresses ways to improve communication skills in dealing with subordinates. The program addresses the following areas which are vital to the success of the operation:

- Coaching and counseling techniques
- Team building skills
- The need for goal and task definition
- Performance appraisals

An outstanding feature of the Forum program is that participants are provided feedback in regard to their handling of case studies. The feedback is given by experts in the above areas.

It is important to realize that today's exceptional manager goes far beyond the traditional business school academic education. Today, an increased emphasis is placed on the psychological aspects of management and the need to develop interpersonal skills. The Situation Management Systems and the Forum Corporation programs were highlighted, but there are several good educational programs which deal with the behavioral characteristics of people. It appears that for at least the next decade, the communications aspect of leadership will be of paramount importance.

LEADERSHIP FACTORS IN R&D

The R&D manager of the future must not only be technically well qualified but oriented to people as well. He must exhibit a greater professionalism in the demanding years ahead and be skilled in behavioral science as well as knowledge in the application of basic business principles. Tomorrow's executives must emphasize that skilled responsible management and superior productivity must go together in successful organizations. These organizations will stimulate productivity in this critical growth ingredient by treating their people as partners. Their respect for the individual will be reflected in the completeness of their people orientation methods that blend effectively with their high expectations for performance.

These basics extend beyond the R&D groups, with many of the factors applicable to all organizational levels and types. Indeed, similar conclusions will be developed concerning the motivational impact of organizational structure, culture, and managerial leadership on productivity.

Specific approaches using the above productivity improvement factors are unique to each leader and organization. The achievements for each leader, group, and organization are dependent upon the perseverence, awareness, ingenuity, and propensity for action demonstrated by each.

Interactions of several variables affect the level of productivity. Typically included are personal factors, job-related factors, and project-related items such as cost and performance. R&D productivity effectiveness must, of course, be evaluated on its long-term impact on timely release of profitable products.

Quantitative-factor measurement such as profit per R&D dollar, drafting time per drawing, and error rate measure are suited decreasingly as task complexity increases. Qualitative assessments (i.e., product performance throughout its life cycle or an organization's ability to win proposal competitions) increase in effectiveness with increasing task complexity. Their proper selection and fairness in their application are certainly critical to the overall respect level between the company management and their employees. The best R&D leaders, managers, or organization will be unable to overcome the unfortunate loss of employee motivation toward productivity if its measurement or assessment methods are judged unfair by those employees.

Critical to the success of the R&D activity is the application by its leader of the key operational functions of productive long- and short-range planning, organization, staffing, and directional control. R&D leadership which actively solicits participation by subordinates in planning for the best uses of resources toward a unified effort, with fair evaluation methods and accounting for future risks, will more likely develop a true sense of ownership by their staff and a stronger commitment to those goals.

It is equally important that the organizational structure be based upon the most effective grouping of functions perceived as important to getting the job done in the most productive manner. Modern managers support the principle that the organization should never be an end in itself but simply an alignment of resources for their best possible utilization. Responsibility must be clearly identified with decision making, authority, and accountability for performance.

Behavioralists teach that no one structure best meets the trend toward shorter product life cycles and R&D's time constraints to effectively shape product features toward ever-tighter market segments. The need to achieve a balance between the technical and social requirements requires the compromise between traditional and the behaviorist concept. The resulting matrix type of organizational structure provides advantages such as a stimulus for productive effort and technological development, simultaneous effective management of a variety of product lines, and the necessary improved communication across the organization with cross-fertilization of technology between programs. While it also minimizes technical obsolescence, its successful use requires a careful blend in responsibility control between the base technology managers and the project managers.

An effective R&D leadership talent base may initially be recognizable in the "embryo" stage by the candidate's ability to elicit strong positive reactions from peers as the candidate begins to fulfill the need for growth. His use for technical support or for solution logic verification by his peers is easily recognized—and often indicates he is a "listener."

No two leaders have the same balance in technical and people-handling skills and therefore leadership style. Leadership styles should be unique—and the effectiveness of a particular style will vary with the group and the task type or complexity, etc. While most managers develop some leadership ability, true leadership development is a lifelong process that has been influenced to a significant degree by our childhood environment.

An organization can, however, act as the catalyst to enhance and develop inherent potential for leadership. Selections for such development should have shown promise and a motivation toward leadership. The organization can then provide the proper climate, opportunity, and challenge—and, of course, the incentive for their development of the necessary skills.

LEADERSHIP IN A PROJECT ENVIRONMENT

Project managers are often selected or not selected because of their leadership styles. The most common reason for not selecting an individual is his inability to balance the technical and managerial project functions. Wilemon and Cicero have defined four characteristics of this type of situation:[48]

1. The greater the project manager's technical expertise, the higher the propensity that he will overly involve himself in the technical details of the project.
2. The greater the project manager's difficulty in delegating technical task responsibilities, the more likely it is that he will overly involve himself in the technical details of the project (depending upon his expertise to do so).
3. The greater the project manager's interest in the technical details of the project, the more likely it is that he will defend the project manager's role as one of a technical specialist.
4. The lower the project manager's technical expertise, the more likely it is that he will overstress the nontechnical project functions (administrative functions).

[48] D. L. Wilemon and John P. Cicero, "The Project Manager: Anomalies and Ambiguities," *Academy of Management Journal,* **13** (1970):269–282.

There have been several surveys to determine what leadership techniques are best. The following are the results of a survey by Richard Hodgetts:[49]

- Human-relations-oriented leadership techniques
  - "The project manager must make all the team members feel that their efforts are important and have a direct effect on the outcome of the program."
  - "The project manager must educate the team concerning what is to be done and how important its role is."
  - "Provide credit to project participants."
  - "Project members must be given recognition and prestige of appointment."
  - "Make the team members feel and believe that they play a vital part in the success (or failure) of the team."
  - "By working extremely close with my team I believe that one can win a project loyalty while to a large extent minimizing the frequency of authority-gap problems."
  - "I believe that a great motivation can be created just by knowing the people in a personal sense. I know many of the line people better than their own supervisor does. In addition, I try to make them understand that they are an indispensable part of the team."
  - "I would consider the most important technique in overcoming the authority-gap to be understanding as much as possible the needs of the individuals with whom you are dealing and over whom you have no direct authority."
- Formal-authority-oriented leadership techniques
  - "Point out how great the loss will be if cooperation is not forthcoming.
  - "Put all authority in functional statements."
  - "Apply pressure beginning with a tactful approach and minimum application warranted by the situation and then increasing it."
  - "Threaten to precipitate high-level intervention and do it if necessary."
  - "Convince the members that what is good for the company is good for them."
  - "Place authority on full-time assigned people in the operating division to get the necessary work done."
  - "Maintain control over expenditures."

[49] Richard M. Hodgetts, "Leadership Techniques in Project Organizations," *Academy of Management Journal*, **11** (1968):211–219.

- "Utilize implicit threat of going to general management for resolution."
- "It is most important that the team members recognize that the project manager has the charter to direct the project."

In the project environment there exists a definite impact due to leadership emphasis. The leadership emphasis is best seen by employee contributions, organizational order, employee performance, and the project manager's performance.

- Contributions from people
  - A good project manager encourages active cooperation and responsible participation. The result is that both good and bad information is contributed freely.
  - A poor project manager maintains an atmosphere of passive resistance with only responsive participation. This results in information being withheld.
- Organizational order
  - A good project manager develops policy and encourages acceptance. A low price is paid for contributions.
  - A poor project manager goes beyond policies and attempts to develop procedures and measurements. A high price is normally paid for contributions.
- Employee performance
  - A good project manager keeps people informed and satisfied (if possible) by aligning motives with objectives. Positive thinking and cooperation are encouraged. A good project manager is willing to give more responsibility to those willing to accept it.
  - A poor project manager keeps people uninformed, frustrated, defensive, and negative. Motives are aligned with incentives rather than objectives. The poor project manager develops a "stay out of trouble" atmosphere.
- Performance of the project manager
  - A good project manager assumes that employee misunderstandings can and will occur, and therefore blames himself. A good project manager constantly attempts to improve and be more communicative. He relies heavily on moral persuasion.
  - A poor project manager assumes that employees are unwilling to cooperate and therefore blames subordinates. The poor project manager demands more through authoritarian attitudes and relies heavily on material incentives.

Management emphasis also impacts the organization. The following four categories show this management emphasis resulting for both good and poor project management:

- Management problem solving
  - A good project manager performs his own problem solving at the level for which he is responsible through delegation of problem-solving responsibilities.
  - A poor project manager will do subordinate problem solving in known areas. For areas which he does not know, he requires that his approval be given prior to idea implementation.
- Organizational order
  - A good project manager develops, maintains, and uses a single integrated management system in which authority and responsibility are delegated to the subordinates. In addition, he knows that occasional slippages and overruns will occur, and simply tries to minimize their effect.
  - A poor project manager delegates as little authority and responsibility as possible, and runs the risk of continual slippages and overruns. A poor project manager maintains two management information systems: one informal system for himself and one formal (eyewash) system simply to impress his superiors.
- Performance of people
  - A good project manager finds that subordinates willingly accept responsibility, are decisive in attitude toward the project, and are satisfied.
  - A poor project manager finds that his subordinates are reluctant to accept responsibility, are indecisive in their actions, and seem frustrated.
- Performance of the project manager
  - A good project manager assumes that his key people can "run the show." He exhibits confidence in those individuals working in areas in which he has no expertise, and exhibits patience with people working in areas where he has a familiarity. A good project manager is never too busy to help his people solve personal or professional problems.
  - A poor project manager considers himself indispensable, is overly cautious with work performed in unfamiliar areas, and becomes overly interested in work he knows. A poor project manager is always tied up in meetings.

SUMMARY

Leadership is an important subject for the team members to understand. Engineering teams usually have an appointed formal leader; yet during the course of a team's work the leadership role will gravitate to the team member who has the expertise to lead the team in a particular problem or opportunity.

We reviewed the definition and evolution of the theory of leadership, the types and functions of leaders, leadership traits and styles, and "participative" leadership. Then we reviewed the cultural impact on leadership and the demands placed on leaders by contemporary organizations. We addressed the appropriate leadership style for project management, life cycle leadership, and leadership factors in research and development.

In Chapter 4 we review the concept and practice of motivation.

BIBLIOGRAPHY

Allport, G. *Social Psychology*. Boston: Houghton Mifflin, 1924.

Argyris, C. *Interpersonal Competence and Organizational Effectiveness*. Homewood, Ill.: Irwin-Dorsey, 1962.

Back, K. *Beyond Words*. New York: Wiley, 1969.

Benne, K. D. *A Conception of Authority*. New York: Teachers College, Columbia University, 1943.

Bennis, W. *American Bureaucracy*. Chicago: Aldine, 1970.

Bernard, L. L. *An Introduction to Social Psychology*, New York: Holt, Rinehart and Winston, 1926.

Blake, R. R., and J. S. Mouton. *The New Managerial Grid*. Houston: Gulf Publishing, 1978.

Blau, P. *Exchange and Power in Social Life*. New York: Wiley, 1964.

Certo, S. C. *Principles of Modern Management*. Dubuque, Iowa: Brown, 1983.

Cooler, C. H. *Human Nature and the Social Order*. New York: Scribners, 1902.

Dahl, R. A. *Who Governs?* New Haven: Yale University Press, 1961.

Davis, R. C. *The Fundamentals of Top Management*. New York: Harper & Row, 1951.

Dowling, J., and J. Pfeffer. *Pacific Sociological Review*, vol. 1, 1975.

Etzioni, A. *Modern Organizations*. New York: Free Press, 1964.

Etzioni, A. *Complex Organizations*. New York: Free Press, 1972.

Fiedler, F. E. *A Contingency Model of Leadership Effectiveness*. New York: Academic Press, 1964.

Fiedler, Fred. "Engineer the Job to Fit the Manager." *Harvard Business Review*, **43**(5), September–October 1965.

French, J. R. P., and B. Raven. *The Bases of Social Power*. Ann Arbor: University of Michigan Institute for Social Research, 1959.

Gross, E. "Dimensions of Leadership," *Personnel Journal*, vol. 40, 1961.

Halpin, A. W., and B. J. Winer. *A Factorial Study of the Leader Behavior Description*. Columbus, Ohio: Ohio State University, Bureau of Business Research, 1957.

Hersey, Paul, and Kenneth Blanchard. *Management of Organizational Behavior*. Englewood Cliffs, N.J.: Prentice-Hall, 1979.

Hodgetts, Richard M. "Leadership Techniques in Project Organizations." *Academy of Management Journal*, vol. 11, 1968.

Hollander, E. *Leadership Dynamics: A Practical Guide to Effective Relationships*. New York: Free Press, 1978.

Hunter, F. *Community Power Structure*. Chapel Hill: University of North Carolina Press, 1953.

Jennings, E. E. *An Anatomy of Leadership: Princes, Herdes, and Supermen*. New York: Harper & Row, 1960.

Julian, J. W. *Leader and Group Behavior as Correlates of Adjustment and Performance in Negotiation Groups*. Dissertation Abstract, 1965.

Kumar, P. *Psychology Studies*, vol. 10, No. 2, 1965.

Lawrence, P., and J. Lorsch. *Organizations and Environment*. Cambridge, Mass.: Harvard University Press, 1967.

MacKenzie, R. A. "The Management Process in 3-D." *Harvard Business Review*, vol. 47, 1969.

Mahoney, T. A., T. H. Jerdee, and S. T. Carroll. "The Job(s) of Management." *Industrial Relations*, 1965.

Merton, R. *Social Theory and Social Structure*. New York: Free Press, 1967.

Miller, J. A. *A Hierarchical Structure of Leadership Behavior*. Rochester, N.Y.: University of Rochester Management Research Center, Technical Report No. 66, 1973.

Mintzberg, H. *The Nature of Managerial Work*. New York: Harper & Row, 1973.

Newstetter, W. I., M. J. Feldstein, and T. M. Newcomb. *Group Adjustment: A Study in Experimental Sociology*. Cleveland, Ohio: Western Reserve University, 1938.

Peters, T. J., and R. H. Waterman, Jr. *In Search of Excellence*. New York: Harper & Row, 1982.

Stodgill, R. M. "Personal Factors Associated with Leadership: A Survey of the Literature." *Journal of Psychology*, 1948.

Tannenbaum, Robert, and Warren H. Schmidt. "How to Choose a Leadership Pattern." *Harvard Business Review*, **51**(3), May–June 1973.

Vroom, V. H., and P. W. Yetton. *Leadership and Decision-Making*." Pittsburgh: University of Pittsburgh Press, 1973.

Wilemon, D. L., and John P. Cicero. "The Project Manager: Anomalies and Ambiguities." *Academy of Management Journal*, vol. 13, 1970.

4
MOTIVATION

MOTIVATING PEOPLE

Management has been defined as getting things done through other people. Inherent in this basic concept of being a manager is motivating people. Motivating people to do what they may not want to do. Motivating and challenging those who want to do what is expected. Fortunately or unfortunately, management is an art and a science and thus no exact formulas exist which will magically instill into employees the desire to achieve or excel beyond the normal expectation.

As noted by Erwin S. Stanton in the March 1983 issue of the *Personnel Journal,*

> For the past 20 years, numerous studies, research reports, and well-formulated theories have attempted to explain the complex behavior of people at work and, more importantly, to gain the practicing manager some rather specific prescriptions as to the best way to manage employees for optimized productivity.[1]

It is the purpose of this chapter to examine these theories and provide a framework in which the manager can apply them in the work environment. For it is appropriately noted by Robert C. Beck in his book, *Motivation Theories and Principles,* that studying the reaction of people to various stimuli is an important foundation from which the manager can develop his strategy to get work done through others. He states,

> . . . if we could not predict the behavior of others, social chaos would arise. Not knowing how others would respond, we would be at a loss as to how to act toward other people. But we do manage to make pretty good guesses about other people's behavior, and society does manage to stay more or less glued together. Indeed, if someone behaves in completely unpredictable ways do we not usually say he is "crazy"?[2]

[1] Erwin S. Stanton, *Personnel Journal,* March 1983.
[2] Robert C. Beck, *Motivation Theories and Principles* (Englewood Cliffs, N.J.: Prentice-Hall, 1978), pp. 8,9.

THE NATURE OF MOTIVATION

Saul W. Gellerman in his article, "The Meaning of Motivation: The Self-Concept," describes the ultimate goal of motivation as

> . . . to make the self-concept real: to live in a manner that is appropriate to one's preferred role, to be treated in a manner that corresponds to one's preferred rank, and to be rewarded in a manner that reflects one's estimate of his own abilities.[3]

Gellerman suggests that motivation is an internal force that compels the individual to seek out the behavior he believes will attain for him the status, rank, recognition, and salary which will match his concept of the ideal he wishes to be. Thus, what motivates one employee will not necessarily motivate another. For the manager this means a tailoring of his interactions with his employees to address those motivating factors unique to each of his employees. This is not an easy task, but the signals issued by the employees are a good indicator of what motivators are considered important to them.

Beck attempts to describe the common understanding of motivation as follows:

> For the man on the street (mythical though he may be), the relationship of mind and body is probably clear. The "official doctrine" . . . is that "body" is physical, material, limited in space, time and size, and objectively observable. "Mind," on the other hand, is just the opposite of all those qualities. It is subjective, directly known only to the individual possessing it, unlimited in physical dimension, and is, perhaps, everlasting. . . . As the man on the street might view it, he is aware of his circumstances and his own feelings and ideas. Faced with several possible actions, he consciously and freely *wills* himself to do this or that.[4] (Theories of Plato and René Descartes)

The employee sees his actions as a simple process of decision making where the "mind" directs the body to perform certain actions. Thus it is that motivation is truly concerned with choice, persistence and vigor toward goals established by the individual (consciously or uncon-

---

[3] Saul W. Gellerman, "The Meaning of Motivation: The Self-Concept," in *Organizational Behavior: Readings and Exercises,* 6th ed., by Keith Davis and John W. Newstrom (New York: McGraw-Hill, 1981), p. 78.

[4] Beck, op. cit., p. 3.

sciously). It is this persistence toward a goal that makes motivation a significant factor in employee relations.[5]

T. H. Newcomb, a noted psychologist, writes that in goal-directed behavior:

> "Motive," like the non-technical terms "wants" and "desire," is a word which points both inward and outward. Such terms refer both to an inner state of dissatisfaction (or unrest, or tension, or disequilibrium) and to something in the environment (like food, mother's presence, or the solution to a puzzle) which serves to remove the state of dissatisfaction . . . [he says that an organism is motivated] when—and only when—it is characterized both by a state of drive and by a direction of behavior towards some goal which is selected in preference to all other possible goals. Motive, then, is a concept which joins together drive and goal.[6]

Here Newcomb raises the relationship of motivation between the internal workings of the individual's mind and the external environment in which he functions. There is, then, a correlation between what motivates a person and the working conditions, cultures, and business practices in which the employee exists.

In discussing the subject of choice of behaviors, Beck notes that individuals choose actions which are expected to have "desirable outcomes" and avoid those actions which will have "unpleasant" or adverse outcomes.[7] Thus the employee draws conclusions about the outcomes of potential actions in the process of motivating himself to perform. Again, based on goals, any action is chosen because it is perceived by the individual to enable positive results which will help him to attain his goals.

In evaluating the filtering process that the employee utilizes to evaluate the desirability of an outcome associated to an action, Keith Davis in his article, "Low Productivity? Try Improving the Social Environments," notes that there are two types of motivation: Micromotivation and macromotivation.[8] Micromotivation deals with working conditions within a single company, whereas macromotivation relates to the social environment outside the company. Davis notes that the employee's social environment determines

[5] Ibid., p. 24.

[6] R. S. Peters, *Concepts of Motivation* (New York: Humanities Press, 1960), p. 39.

[7] Beck, op. cit., p. 25.

[8] Keith Davis, "Low Productivity? Try Improving the Social Environment," in *Organizational Behavior: Readings and Exercises*, 6th ed., by Keith Davis and John W. Newstrom (New York: McGraw-Hill, 1981), p. 79.

- Their attitudes toward work
- Their feelings about work conditions
- Their response to incentives
- Their expectations about supervision
- Most of their other job responses

It is the external environment to the company which strongly influences the employee's outlook or attitude toward work and the work atmosphere.

A person's abilities, that is, his knowledge and skills, combine with his basic desires and attitudes to effect his performance. It is important not to forget the fact that motivation is not the *only* ingredient to the performance of an individual; knowledge of the facts needed to do the job plus basic skills are also necessary. Thus, in evaluating performance problems, the manager should be aware of all factors which result in the total performance of the employee. Ability problems tend to be easily identified and corrective action (training) is usually clear-cut. However, motivation is often hard to uncover and can change frequently, and the presence of more than one motivating factor can occur, thus making the supervisor's task of selecting corrective action difficult.

Before launching into a discussion of motivation and its causes and methods of influencing, let us focus on what a motivated employee looks like. David McClelland has described several characteristics of an achiever:

—He likes situations in which he takes personal responsibility for finding solutions to problems.
—. . . [he has a] tendency to set moderate achievement goals and to take "calculated risks."
—. . . [he] wants concrete feedback as to how well he is doing.[9]

The motivated employee is a self-starter. He is dedicated to performing all tasks to the utmost of his abilities. He sees the accomplishment of tasks as instrumental in the attainment of goals, and he is cooperative with the management philosophy to the extent reasonably possible.

Thus, motivation is a basic ingredient of performance and is composed of those elements which instill in a person the desires to achieve goals established by the individual as important. Motivation is shaped by many

---

[9] David McClelland, "Characteristics of Achievers," in *Organizational Behavior: Readings and Exercises*, 6th ed., by Keith Davis and John W. Newstrom (New York: McGraw-Hill, 1981), p. 44.

external social and educational experiences which the individual has internalized into goals. Thus, if the manager can identify the goals of an employee as well as the factors which the individual believes will enable him to achieve those goals, then the manager has a better chance at providing meaningful tasks which address and capitalize on the employee's desires.

THE BASICS OF ORGANIZATIONAL BEHAVIOR

The study of organizational behavior sprang out of the study of individual behavior. According to Herbert G. Hicks, "organizational behavior is individual behavior that occurs in association with others."[10] Because most people live, work, and recreate in the presence of others, organizational behavior would seem to have an important impact on today's society. As Hicks notes, "research almost always has shown that when persons undertake activities in relation to others, or even simply in the presence of others, certain patterns of behavior—called organizational behavior—can be described.

Organizational behavior includes such topics as

- Cooperation
- Competition
- Social organizations
- Institutions
- Values
- Languages
- Kinship
- Technology
- Social and cultural dimensions

In dealing with these subjects, one can obtain a better appreciation for the big picture. Thus, a systems approach to the environment of an individual can be undertaken.

Hawthorne Studies

The beginnings of organizational behavior were rooted in the work of Elton Maye, F. J. Roethlisberger, and William Dickson in their work

[10] Herbert G. Hicks, *The Management of Organizations: A Systems and Human Resources Approach*, 2nd ed. (New York: McGraw-Hill, 1972), p. 110.

which was to be later referred to as the Hawthorne Studies. Jay Lorsch summarizes the results as follows:

> . . . it was learned that worker behavior is the result of a complex system of forces including the personalities of the workers, the nature of their jobs, and the formal measurement and reward practices of the organization. Workers behave in ways that management does not intend, not because they are irresponsible or lazy but because they need to cope with their work situation in a way that is satisfying and meaningful to them.[11]

This discovery has led to the analysis of performance problems in light of the personality, job, and rewards systems in place in the specific situation.

Basic Forces Affecting Behavior

Jay W. Lorsch in his article, "Situational Theories of Behavior," defines the basic forces shaping behavior and their interrelationship as shown below:[12]

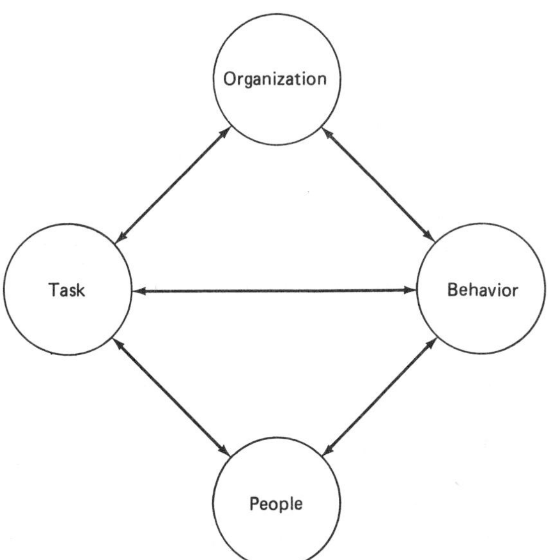

All performance is a by-product of the interrelationships of these factors.

[11] Jay W. Lorsch, "Situational Theories of Behavior," in *Organizational Behavior: Readings and Exercises*. 6th ed., by Keith Davis and John W. Newstrom (New York: McGraw-Hill, 1981), p. 32.
[12] Ibid., p. 35.

HIERARCHY OF NEEDS—MASLOW

A. H. Maslow has contributed to the modern theories of the psychology of motivation. Maslow developed his "hierarchy of needs" as an explanation of the variations between individuals in terms of what tends to motivate them. Maslow contended that

- Human behavior is determined by unsatisfied needs.
- Human needs exist in a hierarchy of importance.
- Higher needs differ from lower needs in that they are never completely satisfied.
- Motivation comes from within an individual.

Maslow's hierarchy of needs are as follows:

SELF-ACTUALIZATION
SELF-ESTEEM
SOCIAL
SECURITY
PHYSIOLOGICAL NEEDS

Maslow contended that humans move up the scale based on satisfying the previous level's needs. The first level is that of the basic or physiological needs, namely food, water, clothing, shelter, sleep, and sexual satisfaction. Simply speaking, man's primal desire to satisfy these basic needs motivates him to do a good job. However, once a need becomes satisfied, man is no longer motivated unless there is a lower-level need which motivates him further. Fulfilled needs are not motivators.

Once an employee has fulfilled his physiological needs, he then turns to the next lower need, safety. Safety needs include economic security and protection from harm, disease, and violence. Safety needs must be considered on projects which may include handling of dangerous materials or anything else which could produce bodily harm. Safety can also include security. It is important that project managers realize this because project managers may find that as a project nears termination, functional employees are more interested in finding a new role for themselves than in giving their best to the current situation.

The next level contains the social needs. This includes love, belonging, togetherness, approval, and group membership. It is at this level where the informal organization plays a dominant role. Many may refuse promotions to project management (as project managers, project office personnel, or functional representatives) because they fear that they will lose their "membership" in the informal organization. This problem can occur even on short-duration projects.

The two lowest needs are esteem and self-actualization. The esteem need includes self-esteem (self-respect), reputation, the esteem of others, recognition, and self-confidence. Highly technical professionals are often not happy unless esteem needs are fulfilled. For example, many engineers strive to publish and invent as a means of satisfying these needs. These individuals often refuse promotions to management because they believe that they cannot satisfy esteem needs in this position. Being called a manager does not carry as much importance as being considered an expert in one's field by one's peers. The lowest need, self-actualization, includes doing what one can do best, desiring to utilize one's potential, full realization of one's potential, constant self-development, and a desire to be truly creative. Many good team managers find this level the most important and consider each new project a challenge by which they can achieve self-actualization.

Project managers must motivate these temporarily assigned individuals by appealing to their desires to fulfill the lowest two levels. Of course, the motivation process should not be developed by making promises the team manager knows cannot be met. Team managers must motivate by providing

- A feeling of pride or satisfaction for one's ego
- Security of opportunity
- Security of approval
- Security of advancement (if possible)
- Security of promotion (if possible)
- Security of recognition
- A means for doing a better job, not a means to keep a job

Understanding professional needs is an important factor in helping people realize their true potential. Consider the following list:

- Interest and challenging work
- Professionally stimulating work environment
- Professional growth
- Overall leadership (ability to lead)
- Tangible rewards
- Technical expertise (within team)
- Management assistance in problem solving
- Clearly defined objectives
- Proper management control

- Job security
- Senior management support
- Good interpersonal relations
- Proper planning
- Clear role definition
- Open communications
- Minimizing changes

Motivating employees so that they have a feeling of security on the job is not easy, especially since the project has finite lifetime. Specific methods for producing security in a project environment include

- Letting your employees know why they are where they are
- Making the individuals feel that they belong where they are
- Placing the individuals in the positions for which they are properly trained
- Letting your employees know how their efforts fit into the big picture

Since team managers cannot motivate by promising material gains, they must appeal to each person's pride. The characteristics of proper motivation include the following:

- Adopt a positive attitude.
- Do not criticize management.
- Do not make promises that cannot be kept.
- Circulate customer reports.
- Give each person the attention he requires.

There are several ways of motivating team personnel. Some effective ways include

- Giving assignments that provide challenges
- Clearly defining performance expectations
- Giving proper criticism as well as credit
- Giving honest appraisals
- Providing a good working atmosphere
- Developing a team attitude
- Providing a proper direction (even if Theory Y)

The following are some key success factors for effective task team action:

- Disavowal of any acts or manifestations of autocratic management
- Recognition of what people need to develop and feel happy in their professional lives
- Ongoing development of a less confrontational management style
- Clear reporting relationships to the formal organizational structure described in charters, policies, and procedures
- Statement of the objective of the teams in such a manner that there can be no doubt as to the decision role of the team
- Existence of an organizational steering committee which serves to couple task team work and overall organizational purposes
- Establishment of authority, responsibility, and accountability of the team
- Promotion of a cultural ambience in the organization that provides the task team's visibility and acceptance by all members of the organization
- Ongoing communication within the organization reinforcing the belief that the task team would share decisions, results, and rewards in the management of the organization
- Use of task teams at all levels of the organization, avoiding any hint that such teams are to be used only at lower levels of the organization
- Setting of the objectives of the sources and the environment so people on the teams can attain these objectives
- Finally, the keeping open of management doors and communication channels so that people are given the opportunity to talk things out

MOTIVATIONAL HYGIENE THEORY—HERZBERG

Frederick Herzberg built on the work of Maslow in his "motivational hygiene theory." In his book, *The Managerial Choice,* Herzberg notes that Maslow was wrong because lower-order needs never get satisfied. He maintained, "the factors that are involved in producing job satisfaction (and motivation) are separate and distinct from the factors that lead to job dissatisfaction." He further defines two sets of needs:

1. Built-in drives to avoid pain, plus all learned drives
2. The ability to achieve[13]

---

[13] Frederick Herzberg, *The Managerial Choice* (Irwin, Ill.: Dow Jones, 1976), p. 58.

Basically, Herzberg's profile of motivators and hygiene factors in an organization are as follows:

Hygiene Job Dissatisfaction	Motivators Job Satisfaction
	Achievement
	Recognition for achievement
	Work itself
	Responsibility
	Advancement
	Growth
Company policy and admin.	
Supervision	
Interpersonal relations	
Working conditions	
Salary[a]	
Status	
Security	

[a] Salary is basically a hygiene factor but often takes on some of the properties of a motivator, with dynamics similar to those of recognition for achievement.

The length of the bars indicate the significance of the item in relationship to other factors. Herzberg's theory includes the recognition that hygiene factors are basically demotivators because their absence will create job dissatisfaction while their satisfaction will not guarantee motivation.

Blake-Mouton Managerial Grid

Drs. Robert R. Blake and Jane S. Mouton developed a theory that managers tend to overemphasize either people or production and thus results would be counterproductive.[14] In their article, "Increasing Productivity Through Behavioral Science," Blake and Mouton propose a grid which seeks to explain the various types of managerial styles in terms of concern for people or production. They see this grid as an important step to helping managers and employees to improve participation.[15]

[14] Rosenbaum, op. cit., p. 13.

[15] Robert R. Blake and Jane S. Mouton, "Increasing Productivity Through Behavioral Science," *Personnel,* May–June 1981, p. 61.

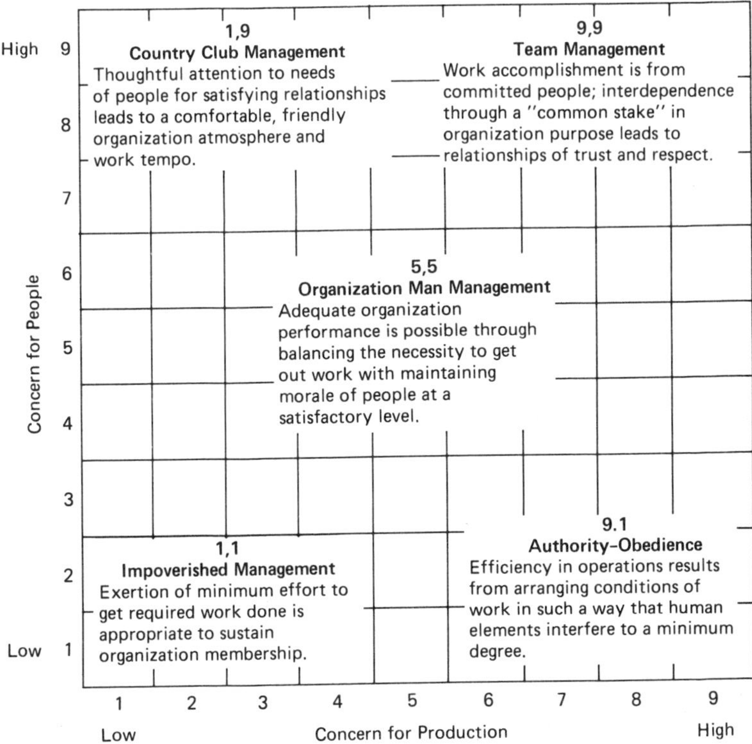

Figure 4-1. The managerial grid. (*Source:* Blake, Robert R., and Jane S. Mouton. *The New Managerial Grid*. Gulf Publishing Company, 1978. Reproduced by permission.)

The grid is displayed in Figure 4-1.[16] When used by managers to discover their basic style, this grid can be a source of further motivational evaluation of which method should apply to certain situations. Matching the right supervisory style to certain work situations can be the needed ingredient to successful productivity.

THEORY X AND THEORY Y—McGREGOR

Douglas McGregor divided managerial attitudes toward employees into two classifications based on the manager's view of people.[17]

---

[16] Ibid., p. 63.

[17] Rosenbaum, op. cit., p. 16.

Theory X—the conventional view of humans engaged in the work situation. According to this view, the average person has an intrinsic aversion to work and will avoid it if possible. The manager must therefore exercise coercion, control, and the threat of punishment to harass workers into even mediocre effort.

Theory Y—the expenditure of both physical and mental effort in work is as natural as play or rest. The average person has an inherent need to be self-motivated and self-controlled, and the shrewd manager is one who knows how to recognize and tap that need so that a worker can learn to accept and seek responsibility.

McGregor's theory in practice shows that many managers are not true Theory X or Theory Y but a compromise. Many managers also are able to adapt one theory or the other in certain situations depending on the circumstances.

PRINCIPLE OF SUPPORTIVE RELATIONSHIPS—LIKERT

Rensis Likert described a principle whereby the supervisor/subordinate relationship is seen as a vital role in the motivational health of a company. Briefly it is stated as follows:

> The leadership and other processes of the organization must be such as to ensure a maximum probability that in all interactions and in all relationships within the organization, each member, in the light of his background, values, desires, and expectations, will view the experience as supportive and one which builds and maintains his sense of personal worth and importance.[18]

Likert points out another big ingredient to motivation when he says, "[it is] the subordinate's perception of the situation, rather than the supervisor's, [which] determines whether or not the experience is supportive."[19] Feeling support from the boss gives the employees self-worth and confidence that what they are doing is correct and appreciated.

---

[18] Rensis Likert, "The Principle of Supportive Relationships," in *Organizational Behavior: Readings and Exercises,* 6th ed., by Keith Davis and John W. Newstrom (New York: McGraw-Hill, 1981). p. 50.

[19] Ibid.

PERSONALITY AND ORGANIZATION—ARGYRIS

Chris Argyris has developed a theme of personality versus the organization which focuses on the range of differences between individuals. The theory states the organization must seek to match

- The needs of the individual
- Requirements (technical and psychological) of the job
- Overall organization climate[20]

This match is critical to the efficient operation of the company.

Argyris notes, "whenever there is an incongruence between the needs of individuals and the requirements of the formal organization, individuals will tend to experience frustration, psychological failure, short-time perspective, and conflict."[21] Argyris further notes that the factors that determine incongruence are

- The lower the employee is in the organization, the less control he has over working conditions.
- The more directive the leadership, the more dependent the employee.[22]

Argyris points to communication and participation as keys to motivation.

THEORY Z

According to William Ouchi, "The secret to Japanese success is not technology, but a special way of managing people, a style that focuses on a strong company philosophy, a distinct Corporate Culture, long-range staff development, and consensus decision making. This style in turn results in lower turnover, increased job commitment, and dramatically higher productivity."[23]

Mr. Ouchi calls this type of dealing with people the "Theory Z approach to management" which suggests that involved employees are the key to increased productivity.[24] We are told that the first lesson of this

---

[20] Chris Argyris, "Personality vs. Organization," in *Organizational Behavior: Readings and Exercises,* 6th ed., by Keith Davis and John W. Newstrom (New York: McGraw-Hill, 1981), pp. 53,54.

[21] Ibid.

[22] Ibid.

[23] William G. Ouchi, *Theory Z: How American Business Can Meet the Japanese Challenge* (Reading, Mass.: Addison-Wesley, 1981).

[24] Ibid., p. 4.

theory is *trust* seconded by subtlety. Subtlety refers to the ever-changing relationships among people, the importance of identifying personalities and matching them to the proper teams to maximize effectiveness. Now let's examine the other key culture values that tie together the three main ingredients of trust, subtlety, and intimacy to form Theory Z.

Lifetime Employment

Lifetime employment is a very important factor since most life activities in Japan are integrated with work. Most Japanese companies hire only once a year, and never fire anyone except for extreme criminal offenses. This policy promotes loyalty to the firm, commitment, and dedication to one's job. It gives employees time to learn how to cooperate rather than compete with one another.

Stable employment that offers an environment of fair equity, challenge, and participation in decision making can overcome voluntary terminations of employees who get better offers elsewhere. Lifetime employment, in short, can foster a two-way commitment between company and employees. They share goals, interest, values, and work together to ensure their survival.

Slow Promotions and Infrequent Evaluations

Since employment is a long-term commitment, Japanese firms are in no hurry to promote or evaluate new employees. Infrequent evaluation avoids the creation of inequality and competitiveness, thus developing a sense of equality which in turn breeds involvement. For example, if for one reason or another a colleague who is not perceived as a superior performer gets promoted ahead of others in his or her group, the self-fulfilling prophecy may come into play as a result. It is as though those left off the promotion list have been told indirectly that they are inferior.

Able employees are given additional responsibility during their formative years while waiting for their time to come. The motivation (and reward) here is not money but membership in working groups which influences behavior. This fact is not new to students of the human relations approach as demonstrated by the Hawthorne Studies. According to Mr. Ouchi, "group memberships in industrial clans have more influence on attitudes, motivation and behavior than does any other social phenomenon."[25] Employees prefer intimate, subtle, and complete evaluation of their peers to external evaluations or rewards.

[25] Ibid., p. 28.

Nonspecialized Career Paths

The policy of nonspecialized career paths forces employees to seek or tap the skills of others and promotes *cooperation* in the process. Generalists rather than specialists result from the Japanese system of career paths. This practice of job rotation helps fill short-run labor needs internally without hiring or firing as those needs change. It may also slow the process of obsolescence.

Collective Decision Making and Responsibility

Collective decision making from the bottom to the top builds consensus, cooperation, and involvement. Collectivism encourages people to work together while building and strengthening their interdependence. Decisions become secondary to commitments and information generated in the process. People tend to support decisions made in a participative manner. They may not agree with chosen decisions, but end up supporting them anyhow because they are fairly arrived at. Also the use of collective (shared) responsibility minimizes the risk of one's being blamed and thus fosters cooperation. The risk is spread out rather than concentrated.

APPLYING MOTIVATIONAL THEORIES

Previous sections have identified many of the popular theories of motivation recognized today. Although these theories have been applied successfully in the past, there is no single, universal theory of motivating performance. The purpose of this section is to analyze the more commonly known motivational theories and to suggest successful ways they may be applied by first-line managers.

Probably the most widely known theory of motivation is Maslow's heirarchy of needs. This theory is useful to managers in that it emphasizes the necessity to identify employees' individual needs and to utilize them by linking their satisfaction to effort or performance. For instance, if an employee is highly motivated by security, the manager may suggest that long tenure with the company depends upon achieving a high level of performance. Of, if a need for personal recognition has been identified, the manager may satisfy that need when a high level of performance is evident.

The major drawbacks to Maslow's theories are that, first of all, individual needs of employees are often difficult for the manager to isolate. Furthermore, Maslow's specified hierarchical arrangement can vary from person to person, and more than one need can be dominant in one person at a time. Maslow's theory also does not spell out when a need has been adequately fulfilled. Therefore, although Maslow's proposal is appealing to managers, it is difficult to apply in practice.

Identification of David McClelland's three motivational forces (affiliation, power, and achievement) in employees can be used by the manager in order to stimulate employee motivation. If an individual's behavior is basically affiliation-motivated, he should be given the opportunity to perform tasks within a self-selected group, and his compensation should be based on group rather than individual productivity. Employees with a high power motive should be given authority over others depending on their degree of skill in this area. High "achievement" employees must have moderately difficult goals, responsibility to accomplish these goals, freedom to do the job their way, and sufficient feedback to know where they stand. The manager must remember that more than one of these motives may be present in an individual at one time, and that often a mix of motivators must be used to meet the needs of certain employees.

Herzberg's two-factor theory represents another popular approach to motivation. Herzberg has identified two factors that affect employee job satisfaction, hygienes and motivators. These factors have been adequately described in previous sections. The important point to note here is that improving the hygiene factors is important and should not be ignored, because doing so could increase employee job dissatisfaction and reduce the level of energy necessary to do the job at hand. However, real motivation can only be stimulated by emphasizing the second series of factors, the motivators. One method proposed to accomplish this has been through what is known as "job enrichment." That is, examine the management functions the employee's immediate boss is doing that the employee would like to do and is capable of doing—then, delegate them to him.[26] Another method commonly used to achieve the same result is "management by objectives" (MBO). There are four steps the manager can use to successfully apply MBO. The first step involves manager and employee collaboration on a set of achievable yet challenging goals for the employee to strive toward. The manager then discusses with the employee how these goals will be accomplished. The manager can suggest ideas of his own; however, no attempts to force these ideas upon the employee must be made. Implementation of these goals is the next logical step. Self-direction and self-control by the employees are the keys here; however, care should be taken to encourage employees to change goals if

[26] Brian H. Kleiner, "Integrating Major Motivational Theories," *Journal of Systems Management,* February 1983, pp. 26–29.

necessary. Finally, the evaluation step compares goals with actual results. If goals are met, reward the employee appropriately. If not, discuss reasons for failure. All in all, MBO represents a philosophy of management based upon the assumption that there is real power in people setting their own goals.

The expectancy/valence theory has provided managers with several significant motivational tools. If an employee's motivational level is deemed inadequate, each of the three basic components of the motivational model (E-P expectancy, P-O expectancy, and valence) must be examined to determine the location of the weak spot. If the E-P component is weak, this means that the employee believes his effort is not related to job performance. This could be true if performance is influenced by factors not under the employee's control. However, low self-esteem or self-confidence could be the culprit. In this case, the manager could work on raising the employee's self-image to correct this problem. If the P-O element is weak, it generally indicates that performance is not being adequately rewarded. A manager might turn his attention toward trying outcomes more directly to performance through the use of incentive plans, commissions, merit raises, or merit-based promotions.[27] If the valence component is the weak link in the chain, it indicates that available rewards are not valued by the employees. Managers must take care in this case to examine more closely their employees' relevant and valued outcomes and match them with performance rewards.

The Blake-Mouton managerial grid can be a useful tool to managers in discovering their managerial style and the subsequent effect it has on employee motivation. The ideal manager in the Blake-Mouton study is the 9, 9 team manager, who recognizes the importance of both people and work accomplishment to his effectiveness as a manager. By analyzing his own outcome, a manager can discover his own managerial style and develop ways to come close to the 9, 9 managerial description. Thus, the organizational climate can be improved to stimulate motivation on the part of both employees and management.

Reinforcement theory, while quite controversial, can be used as a motivator if used carefully. Positive reinforcers such as praise, recognition, pay, or promotion can be used to increase the occurrence of desired employee behavior. The manager must, however, be sincere in his use of praise and must use it with discretion rather than over-frequently. Also, outcomes to be reinforced should be chosen with care, since individual differences enter into the motivational framework.

[27] David E. Terpstra, "Theories of Motivation—Borrowing the Best," *Personnel Journal,* June 1979, pp. 376–379.

McGregor's Theory X and Theory Y managerial styles provide an interesting framework for the study of motivation. McGregor's two managerial descriptions have been interpreted by many as labels to apply to appropriate managers. However, McGregor only intended X and Y to be two points on a continuum, with most managers falling somewhere between the two. Nor did he intend to imply that Theory Y managers were the ideal. Rather, Theory Y was developed to point out that the kind of work people did could be important to them and that work itself could be a powerful motivator. There are, in any organization, some people who want more direction and some who want less. Some people are psychologically dependent upon others; others are autonomous and prefer to direct themselves.[28] McGregor's Theory X and Y intends to make us aware of this point, and that it is entirely possible to be a Theory Y type of person who achieves his own personal objectives by helping his organization achieve theirs.

Theory Z attempts to go even further than Theory Y by adapting Japanese organizational behavior techniques to the United States. The specifics of this approach have been discussed previously. The main question to be answered is whether this managerial philosophy has any merit in motivating employees to a greater productivity. While there is some evidence that the Theory Z approach has been successful in isolated cases in the United States, empirical evidence is not great enough to justify full confidence in this approach. There is considerable skepticism as to the feasibility of clanlike organizations operating in this country. Our increased mobility tends to destroy any attempts at increasing the feelings of oneness within an organization. Furthermore, the American self-interest motives on the part of business do not support the development of a selfless organizational commitment on the part of the employees. In addition, there will always be employees who cannot be socialized into the clan way of doing things. Finally, trust between management and employee ranks is the main ingredient of Theory Z management. There is some doubt as to whether egalitarianism is related to trust. Research seems to indicate that trust is more closely tied to hierarchical managerial structures. It is apparent that this approach needs more study before it can be wholeheartedly applied for motivational purposes in this country.

In summarizing the preceding motivational theories, we can make some general recommendations. Any motivational attempt must begin by identifying an individual's needs. Then work experience can be structured to accommodate fulfillment of those needs. Care should be taken in selection

---

[28] Thomas L. Quick, *The Quick Motivation Method* (New York: St. Martin's Press, 1980), pp. 28–29.

of work rewards and outcomes. It is also very important to incorporate goal setting into this framework. Goals should be challenging, yet realistic. Finally, successful attainment of specified goals should be positively reinforced. By adopting bits and pieces of the motivational theories described, managers can achieve a more complete understanding of the process and problems of motivating employees.

WORKING WITH PEOPLE

There are no simple formulas for working with people. The most that can be done at present is to increase understanding and skills so that human realtions in the workplace can be upgraded. This is the basis from which the human relations approach to management springs. The human relations approach means working with people, and it is from this perspective that the subject will be discussed.

The theory behind the human relations approach to management has been previously displayed, but it is important to remember that we can work effectively with people if we are prepared to think about them in human terms. Employee human relations is the study of the employee in human terms at work and represents an effort to take action in operating situations to produce better results. The focus of this study is managerial applications of working with people. A manager is doubly responsible for developing better human relations, since he is responsible not only for his own human relations but also those of the section he manages.

From the viewpoint of the manager, human relations is motivating people in organizations to develop teamwork which effectively fulfills their needs and achieves organizational objectives.[29] There are several significant points in this definition that relate to working with people. First of all, human relations emphasizes people, not economics. The people are in an organizational environment. The key activity in human relations is motivating people—that is, helping people release the inner drives they already have. The direction of motivation is toward teamwork, requiring both coordination of work and cooperation of the people involved. Human relations through teamwork seeks both need fulfillment and the meeting of organizational objectives. People seek their own personal need fulfillment through attainment of organizational objectives; the organization is responsible for providing a climate conducive to meeting these needs.

There is a general framework of the individual need structure which managers must understand in order to effectively work with their people.

---

[29] Keith Davis, *Human Relations at Work* (New York: McGraw-Hill, 1967), pp. 2–20.

Each person has a need structure which consists of intangible as well as tangible needs. This need structure is conditioned by the whole system in which the individual lives. To the extent that an organizational environment can be created which supports fulfillment of these needs, he will be motivated toward personal growth and full contribution of his energy. All people in the organization work toward maintaining this environment, but management performs the most important function: working both with the internal organizational environment and coordinating outside influences with the internal structure. This approach recognizes that people are the source of improvement in organizations, that through growth of its people an organization improves performance, and that through improvement of its organizations a society improves.

A common misconception about human relations is that it is just common sense. Surely there is some truth to this statement, but then why is there such a lack of human relations evident in everyday personal interactions in the workplace? Human relations requires the learning of technical knowledge about human behavior, development of social skills, development of analytical frameworks, and research into conditions affecting each situation.[30] Study is required to develop good interpersonal skills.

Managers cannot mistake socializing with people for working effectively with them. Human relations does not mean liking people, but rather developing constructive working relationships with the organization. Its purpose is to get the job done.

Along the same lines, it is not correct to assume that all conflict is bad in an organization. Only destructive conflict harms the organizational setting—the remainder can be channeled into productive results.

Human relations must not be considered a weakening agent to managerial authority. Quite the contrary, its real aim is to integrate the employees so that their ideas contribute to corporate goals. This, in turn, enables necessary authority to be more effective with people.

Certain fundamental concepts to be applied to the theory of human relations revolve around the nature of humans and the nature of organizations. Managers must understand these concepts in order to work effectively with people.

The nature of humans embodies four basic assumptions. First, each person is an individual and hence must be seen as such by the manager. Second, a whole person, not just a portion, works in an organization. A person's life, for instance, will affect his work performance. Third, motivation exists only to fulfill a person's needs as he sees them. Therefore management can control motivation only to the extent of the employee's

---

[30] Ibid.

point of view. Fourth is the factor of human dignity. People have a right to be treated differently from other factors of production.

The nature of organizations encompasses two principles: (1) They are social systems, and (2) they are formed on the basis of mutual interest. The idea of a social system refers to the fact that both management and personnel are united in a common goal of providing goods and/or services to a community. Only through shared goals can each side hope to attain their personal goals.

Finally, human relations is not just for nonmanagement employees. Human problems can occur at any level, and therefore human relations can and should be applied at all levels of the organization.

Working with people means creating an environment conducive to maintaining all the principles brought out above. The human relations approach helps us achieve this objective.

MANAGING CREATIVITY

Most companies have creative persons within the work force. Many psychological studies have indicated that all people are creative to some degree.[31] With this vast exposure that organizations and managers have to creative persons, it is important that the line, staff, and executive management levels deal with the ways in which creativity may be motivated. The first task necessary is to define and identify the truly "creative" individuals.

Defining the Creative Individual

The less defined and more vague the job description, the greater the likelihood that the person holding that position is creative. Creative persons have creative temperaments and many, if not most, are not "company" people—they have little sense of loyalty to the organization. Many creative persons have a strong need for individual praise and recognition for their efforts.[32]

Creative individuals provide great benefits to an organization. The stimulation of new ideas and products are two of the most noted contributions. In addition, the person who possesses creative talents tends to be adaptable to various situations. These traits aid the organization in its attempts to stay near the state of the art in both the manufacturing and

---

[31] Herbert G. Hicks and W. Jack Duncan, *The Management of Organizations* (New York: McGraw-Hill, 1972), p. 204.
[32] "The Creative Temperament," *Management Review,* **71** (July 1982):6.

administrative processes and human resources utilization. Creativity is the first step in the process of change. For individual applications, the creative individual tends to be project-oriented. These benefits multiply when projects involve intense interaction. Creative individuals tend to have frequent periods of intense and furious activity, leading to project task achievements. It is important that supervisors be able to understand, identify, and utilize the creative individuals within their areas. In order to obtain the benefits, the line supervisors and managers must be able to motivate these individuals.

Motivating the Creative Individual

Just as each individual has different needs, the creative individual has needs which, if met in the proper way, can provide motivation in the work environment. Some of these motivating factors are as follows:

- Change in the organizational environment
- A creative work environment
- Faith in the individual's abilities
- Reinforcement of self-image
- Flexibility in the organizational "rules"
- Acknowledgment of individuality

Each of these factors can influence the behavior of the creative individual. If one item was to be chosen from the list (which is certainly not all-inclusive), the most important factor which motivates creative persons is the existence of a creative environment in which to operate.

According to a study at the University of Utah Conference on the Identification of Creative Scientific Talent, five degrees of environment may exist in the organization which can influence creative behavior:

1. Stimulating
2. Supportive
3. Neutral
4. Hostile
5. Destructive

The best climate is a stimulating climate. Any other climate will not serve the needs of the creative individual's talents. Some ways in which the stimulating climate can be created is through the elimination of vertical communication channel restrictions and the existence of creative managers.

Vertical communication channels limit the means and resources of the creative individual. This type of person needs to deal with all aspects of his or her environment, utilizing the resources of each subsystem or group. By removing this type of restriction, the creative individual feels more freedom to interact, allowing creativity to flourish. The existence of creative managers allows the creative individuals' climate to be more open. The only drawback—or safeguard—which must be taken is to induce the creative manager to delegate. Creativity and idea inception are factors which creative individuals wish to control. Creative managers who do not delegate stifle creativity and negate the climate. With the creative climate in place, the line supervisor or manager must apply creative programs to motivate the employees.

Applying Creativity Programs in the Organization

There are many ways to apply motivating techniques to the creative individual. The creative persons must first be identified. Some jobs which typically attract creative persons are engineers, programmers, copywriters, designers.

There are no sure formulas for stimulating and motivating the individual. Michel Corbey of TOCOM, a Dallas manufacturer, manages his employees by informal meetings. Coffee shop, water cooler, and elevator conversations limit vertical channels and formal structures. The dual management system approach to stimulating creative individuals supports the concept of using the creative individuals for technical work and leaving the administrative functions such as budget control to the less creative individuals. In any system of control, there must be room to allow the creative person to try out ideas. Reward systems for creative persons include surprise luncheons and time off. In any event, creative work assignments must be carefully orchestrated. Time must be allowed for the individual to work, study, and be open for creative thought and reflection.

The freedom of this type of atmosphere and loose forms of monitoring lends itself to unique controlling mechanisms. Human resource management systems develop various conceptual frameworks for controlling individuals in the workplace. One technique supported for the creative individuals is a system which provides for scheduled review meetings. These meetings should be designed to limit the analysis of why projects have succeeded or failed and to concentrate on the approach used to address the projects. The meetings should

- Get project participation *after* the project is completed
- Be controlled by creative individuals, even if they were not involved in the project

- Enhance role definitions of the participants
- Be loosely structured, but controlled—leading to frank discussion of the matters at hand

In all cases, the meetings should be judged on the costs versus the benefits derived. Participation should also be mandatory, but with participants varying or rotating to eliminate dominance of specific persons.

MOTIVATING THE PROBLEM EMPLOYEE

Many of the theories and applications of motivation in the business environment are directed toward the "normal" employee. However, there seems to be some correlation between the amount of motivation needed and the performance level of the employee. The employee whose productivity is less usually has to have more motivation from the supervisor. Because of this, this section is included to identify special areas of concern that a supervisor should concentrate on. The following categories will be identified:

- The dissatisfied employee
- The performance problem employee
- The underachiever
- Socially related problems: breaking rules, illegal activities (stealing), alcoholism, drug addiction, etc.

The manager must be able to detect signs of an employee becoming either dissatisfied or distressed. According to H. Levinson, all signs or symptoms include one of these changes in the behavior of the employee:[33]

- The person's usual manner becomes overemphasized.
- Restlessness and agitation will appear with a difficulty in concentration.
- Actions contrary to typical behavior for that person occur.

Once the dissatisfaction has been detected, the manager must make the effort to counsel the individual, bringing the problem to the surface. The following seven steps outline a plan to bring the employee back into a position where the common motivational techniques can again be applied:[34]

---

[33] Bernard L. Rosenbaum, *How to Motivate Today's Workers: Motivational Models for Managers and Supervisors* (New York: McGraw-Hill, 1982), p. 131.
[34] Ibid.

1. In a supportive manner, ask the employee to discuss his or her area of dissatisfaction.
2. Show understanding by actively listening.
3. Discuss the causes of the dissatisfaction in addition to the symptoms.
4. Ask for the employee's suggestions as to how the causes of the dissatisfaction can be resolved and offer yours.
5. Come to an agreement on the steps each of you can take.
6. Indicate that you value the employee and that you want to see the employee succeed.
7. Set a follow-up date.

The last step could be the most important. If maintenance is not a continual process, the same problem could easily surface again. In some cases, the second time it appears it is much more difficult to correct.

The performance problem is more easily detected. The more accountable the person is for his efforts, the more symptoms will show. Meeting standards as a measure of performance should be approached from a relative viewpoint rather than the absolute. In many cases, the standard may not be correct when the employee is in fact making a satisfactory contribution. Rosenbaum identifies these action steps to be taken for a performance problem:[35]

1. Focus on the performance problem, not the employee. The conversation should be centered around the future rather than the past.
2. Ask for the employee's help in solving the problem and discuss both your ideas and the employee's ideas on how to solve it.
3. Come to an agreement on the problem and write down the steps to be taken by each of you.
4. Express your confidence in the employee's ability to correct the problem.
5. Set a follow-up date.
6. Praise the employee at the first sign of improvement in job performance.

With the performance problem, as with any personnel problem, usually the longer the time duration between detection and attempts for correction, the worse the problem will become. This also leads to less effectiveness of the supervisor's intervention into the problem.[36] Poor or unsatis-

---

[35] Rosenbaum, op. cit., p. 113.
[36] Rosenbaum, op. cit., p. 111.

factory performance that is ignored by management could be interpreted as rewarded performance. This also means that such performance will be repeated.

The underachiever could be considered a problem employee only in the fact that his or her true potential is not being utilized by the organization. To foster higher achievement by these individuals, Davidson identifies five methods for increasing performance:[37]

1. Goal setting
2. Establishing formal authority
3. New and appropriate responsibilities
4. Defusing the time bomb
5. Cleaning up your own backyard

Long- and short-term goals should be established with both the employee and the supervisor formally setting them. In addition to the goals, the steps necessary to meet them should be documented so that the employee does not lose sight of them as time progresses. The lack of formal documentation is usually a problem, as the working relationship is not defined. Many times formal authority is not accepted unless it is in written form.

New responsibilities mean new challenges for the employee. If changes occur in a department or section of an organization, the employee not affected by the changes must be given new and challenging tasks, or he or she will feel passed by. By defusing the time bomb, Davidson means that supervisors have to occasionally turn the other way to defuse an employee's attitudinal problems. However, they cannot be allowed to grow to excessive proportions before action is taken. In keeping his own backyard clean, the manager must continue to show positive attitudes and actions. If the supervisor maintains negative thoughts about the employee or his work, these opinions get transmitted to the employee consciously or unconsciously.

With socially related problems, the scope of the problems should be addressed and attempted to be corrected before effort is made toward improving performance. Although the correction of the socially related problems is well beyond the scope of this chapter, such problems nonetheless have to be recognized by the manager. Communication between the two parties has to exist in order to get a start on the problem. In the best interest of all involved, managers who have not been formally trained

---

[37] Richard Davidson, "Motivating the Underachievers," *Supervisory Management,* January 1983, p. 40.

in correcting these types of problems should contact experts in the relative fields, and let them work with the employee.

REWARD SYSTEMS

Reward systems play an integral part in the employee's and, for that matter, management's motivational process. It can be fairly stated that, as past practices in many organizations have shown, organizations which rely only on financial (i.e., monetary) rewards to motivate employees do not achieve the desired results. This does not mean that money itself is not a powerful motivator, because it is. In fact, according to an article in *Management World,*[38] where the employee has (1) a high net preference for money, (2) a perception that money is tied to performance, and (3) a perception that effort leads to performance, a monetary reward system can indeed motivate greater performance.

In motivating employees, reward systems should not only encompass the monetary aspects but other aspects as well. Employee recognition is one of the most powerful ways a manager has of motivating his employees. Individual praise for a job well done costs the company nothing and yet leads to a higher performance level in most employees. According to Frederich Herzberg's research, there are six primary motivators: (1) recognition, (2) achievement, (3) advancement, (4) possibility for growth, (5) the work itself, and (6) responsibility. Of these six primary motivators, Herzberg picks recognition as one of the strongest. It can be said that most supervisors at most companies do not operate at maximum efficiency simply because they do not receive and do not give praise for performance. Listening, recognition, and rewards combine to form a powerful motivation tool.

Supervisors must realize that in addition to giving rewards, special attention must be given to how to achieve the maximum results from reward giving. Jack J. Phillips, in an article in *Supervision,* has listed five principles of effective reward giving:[39]

1. The reward must be timely.
2. The reward must be appropriate for the achievement.
3. The reward must be perceived as being worthwhile.
4. Most rewarding should be done in public.
5. The reward must be given in a sincere manner.

---

[38] Stephen C. Bushardt, Roberto Toso, and M. E. Schnake, "Can Money Motivate," *Management World,* November 1981, pp. 25–44.
[39] Jack J. Phillips, "Rewarding Employees Effectively: Substance and Style," *Supervision,* November 1982, pp. 14–15, 20, 23.

According to Phillips, that should be the *style* of reward giving. The *substance* of reward giving is almost unlimited and, although by all means not exhaustive, the following are some types of rewards which have been used to effectively recognize employee's efforts:[40]

- Promotion, advancement or transfers
- Pay increases based on merit
- Special assignments
- Performance appraisals
- Special training programs
- Commendation memo/personnel report
- Service awards
- Competitive games and contests
- Suggestion systems
- Free meals
- Recognition for perfect attendance
- Special recognition on birthdays, employment anniversaries
- Committee or task force assignments
- Employee benefits administration
- Charity drives
- Group meetings
- Bulletin boards/employee publications
- Special thanks

The list could go on and on, limited only by the supervisor's imagination. The point is that employee recognition, where deserved, is a powerful tool when used effectively for employee motivation.

Supervisors must understand the requirements, accomplishments, and characteristics of an effective reward system. The nature of the rewards must fit within the employee's value system. Finally, employee satisfaction must be compared with the reward system.

The requirements of an effective reward system have been summarized by J. R. Hackman and J. L. Suttle as the following:[41]

- The system must make enough rewards available so that individuals' basic needs are satisfied.
- The reward levels in the organization must compare favorably with those in other organizations.
- The rewards must be distributed in a way that is seen as equitable by the people in the organization.

---

[40] Ibid.

[41] J. R. Hackman and J. L. Suttle, *Improving Life at Work: Behavioral Science Approaches to Organizational Change* (Pacific Palisades, Calif.: Goodyear, 1977), Chapter 3.

- The reward system must deal with organization members as individuals.

With these requirements in place, the reward system should be able to

- Motivate employees to join the organization
- Motivate employees to come to work
- Motivate employees to perform effectively
- Reinforce the organizational structure by indicating the position of different individuals in the organization.

In using the reward system effectively, supervisors must make the employees aware of the criteria needed to be met to be eligible for rewards and the nature of the rewards themselves. The criteria may be in the form of measurable results, past achievements, seniority, or loyalty, or even tied to consumer indexes. As long as the employee is aware of what the criteria are, he will be more motivated to higher performances in that area. The employee should be aware of the nature of the rewards, whether it be in the form of money, fringe benefits, promotions, job assignments, or even job security. These rewards should (1) have importance to the employee, (2) have flexibility as to who receives them, (3) provide esteem and visibility recognition, (4) be of a frequent nature, and (5) be at a low cost to the organization.

When the supervisor wishes to reward employees for high performance, he must keep in mind the fit between the reward and the value placed upon that reward by the individuals. These will often vary due to

- Different cultural backgrounds
- Education
- Age
- Career
- Aspirations of the individual
- Life styles
- Work experience

Hackman and Suttle state, "research has shown that [employee] satisfaction is a complex reaction to a situation and is influenced by the following factors":[42]

- Satisfaction with a reward is a function of both how much is received and how much the individual feels should be received.
- People's feelings of satisfaction are influenced by comparisons with what happens to others.

[42] Ibid.

- Overall job satisfaction is influenced by how satisfied employees are with both the intrinsic and extrinsic rewards they receive from their jobs.
- People differ widely in the rewards they desire and in how important the different rewards are to them.
- Many extrinsic rewards are important and satisfying only because they lead to other rewards.

The identification of those factors in a reward system that motivate employees and the ability to effectively use those factors in the motivational process is the mark of a good manager. Applying the incentives and recognition process wisely will not only help motivate the employees but will satisfy the organizational goals at the same time.

WORK PERFORMANCE AND THE MOTIVATIONAL PROCESS

All supervisors want to motivate their employees to obtain the highest degree of efficiency possible in their work performance. Discussed in this section are some of the ways the supervisor can motivate employees to perform better and why these techniques will often lead to better work performance.

The noted psychologist Frederick Herzberg conducted extensive research in the areas of intrinsic versus extrinsic motivation and job satisfaction. Out of this research, Herzberg developed what he referred to as the motivator-hygiene theory.[43] Essentially, he believed that what motivates employees and what satisfies employees are unrelated to what causes dissatisfaction and nonmotivated employees. A model of the motivator-hygiene theory looks like the following:

SATISFACTION	*NEUTRALITY*	*DISSATISFACTION*
Motivators		*Hygiene*
Job content (the work itself)		Job content (the work environment)
Achievement		Pay
Growth		Fringe benefit
Advancement		Type of supervision
Responsibility		Work conditions
Recognition		

[43] Donal Sanzotta, *Motivational Theories and Applications for Managers* (Amacom, 1977), N.Y., N.Y., pp. 25–27.

Deficiency of the hygiene factors can lead to dissatisfaction and no motivation. Hygiene factors, then, can be said to be prerequisites for satisfaction (and motivation). From this, it can be said that hygiene factors are never motivators, but will tend to decrease motivation if not present.

In addition to making sure that the above factors are present, the manager should keep in mind that different people are motivated to different degrees depending on such factors as

- Type of work
- Work relationships
- Autonomy of the work
- Personality styles
- Financial rewards
- Sense of achievement
- Social status
- Power and influence

Dean R. Spitzer, in an article appearing in *Training,*[44] has compiled a list of the most often used ways of motivating people to perform better:

- Use appropriate methods of reinforcement—tell employees when they do good work.
- Eliminate unnecessary threats and punishments—these are negative factors and encourage avoidance behavior.
- Make sure that accomplishment is adequately recognized—people need to feel important.
- Provide people with flexibility and choice—permit employees to make decisions when possible.
- Provide support when it is needed—asking for help should never be considered a sign of weakness.
- Provide employees with responsibility along with their accountability—giving people appropriate responsibility is always a good motivator.
- Encourage employees to set their own goals—people tend to know their own capabilities and limitations better than anyone else.
- Make sure that employees are aware of how their tasks relate to personal and organizational goals—explain the "big picture" to employees.
- Clarify your expectations and make sure that employees understand

[44] Dean R. Spitzer, "Thirty Ways to Motivate Employees to Perform Better," *Training*, March 1980, pp. 51–52, 55–56.

them—this does away with the frustration of not being sure what is expected.

- Provide an appropriate mix of entrinsic rewards and intrinsic satisfaction—employees need to obtain personal satisfaction in addition to extrinsic rewards.
- Design tasks and environments to be consistent with employee needs—different people need different activities, use good common sense in making the proper "fit."
- Individualize your supervision—to maximize motivation, treat people as individuals.
- Provide immediate and relevant feedback that will help employees improve their performance in the future—inform employees how they might improve their performance with informational feedback.
- Recognize and help eliminate barriers to individual achievement—if barriers are not dealt with, employees may remain underachievers needlessly.
- Exhibit confidence in employees—this results in positive performance.
- Increase the likelihood that employees will experience accomplishment—"nothing succeeds like success."
- Exhibit interest in and knowledge of each individual under your supervision—people need to feel important and personally significant.
- Encourage individuals to participate in making decisions that affect them—people who have no control over their destiny become passive.
- Establish a climate of trust and open communication—threat is one of the obstacles to high motivation.
- Minimize the use of statutory powers—try managing democratically, encouraging input and participation.
- Help individuals to see the integrity, significance, and relevance of their work in terms of organizational output—again, help the employee see the whole picture and not just pieces.
- Listen to and deal effectively with employee complaints—handle problems before they become blown out of proportion.
- Point out improvements in performance, no matter how small—frequent encouragement is useful in the early stages but should be reduced as confidence and proficiency are obtained.
- Demonstrate your own motivation through behavior and attitude—"one picture is worth a thousand words."
- Criticize behavior, not people—avoid at all costs self-fulfilling prophecies.

- Make sure that effort pays off in results—if effort does not result in accomplishment, effort will be withheld.
- Encourage employees to engage in novel and challenging activities—provide opportunities for employees to try new tasks.
- Anxiety is fundamental to motivation, so don't eliminate it completely—some of the time, the best work is done under pressure.
- Don't believe that "liking" is always correlated with positive performance—a task can be intrinsically boring, while the consequences are highly motivating.
- Be concerned with short-term and long-term motivation—people should receive reinforcement in short-term and long-term assignments.

Of all the ways that a supervisor can use to obtain better performance from his employees, one item stands out above the rest. Employee recognition is still the top motivator of employees. Alone it will not work, but with the other necessary factors, it works extremely well.

PERFORMANCE, PRODUCTIVITY, AND JOB SATISFACTION

In a discussion of motivation in the workplace, performance, productivity, and job satisfaction are all interrelated. Certainly front-line supervisors desire to obtain the best performance possible from their subordinates. In order to achieve optimum productivity from the employees, managers must address employee satisfaction.

Measuring employee satisfaction is not always an easy task. There is no yardstick by which satisfaction can be measured. Indeed, as we have seen, different employees require different things and thus their degrees of job satisfaction varies by preference. A supervisor should, however, be able to tell when employee satisfaction is not at a high level. Any one or a combination of the following factors could indicate that a problem with employee job satisfaction exists:[45]

- Productivity
- Grievances
- Morale
- Absenteeism and tardiness
- Labor turnover

---

[45] Claude S. George, Jr., *Management for Business and Industry* (Englewood Cliffs, N.J.: Prentice-Hall, 1970), pp. 324–330.

Productivity can be an indication when there has been a decrease or slowdown over a relatively short period of time. Grievances represent an excellent indicator, not of the number so much as the type. Low employee morale will lead to a decrease in performance. Management should check to see why absenteeism and tardiness are present and thus determine the cause of the unrest. Labor turnover, outside the norm for the industry, could reveal inadequate training, poor supervision, etc.

Once the supervisor determines that there is a problem with employee job satisfaction, his task should be to correct, if possible, the cause of the dissatisfaction, or at the very least, to address the problem. Several ways in which this can be done are

- Communications
- Personnel management
- Counseling
- Committees and meeting

Recognizing that there is a problem is the starting point for the manager. Communications, from and to the employee, will help to develop an understanding of what the problem is. Keeping the employees informed about the company, its services, and its benefits will also increase employee satisfaction. Where at all possible, the supervisor can get to know the employee individually, giving advice and becoming familiar with the problem. Individuals may be placed on committees and put in charge of meetings to increase their participation and feeling of importance. All of these techniques will have the effect of increasing employee satisfaction.

Two theories which come into play when discussing performance, productivity, and job satisfaction are the reinforcement theory and the expectancy theory. Used and understood, these theories can aid the supervisor in increasing not only performance and productivity, but also employee job satisfaction.

The reinforcement theory does just that. It reinforces the employee's performance. Basically it states that motivation comes through the individual's desire to satisfy physical and psychological needs. Management reinforces these needs by rewarding such behavior as

- High-quality work
- High productivity
- Timely reports
- Creative suggestions

The expectancy theory concerns itself with what the employee expects and carries through to job satisfaction. It merely *emphasizes*

- The employee's perceived relationships
- The valence or value of the reward to the employee
- The belief that performance will be rewarded
- The expectancy or perceived relationship between effort and the ability to perform well

In knowing and utilizing these two theories, supervisors will be more able to achieve higher performance and a higher degree of job satisfaction. However, the supervisor must not forget certain performance factors which concern the employees. Such performance factors may be (not all-inclusive by any means) as follows:

- Employees seek opportunities to discuss their performance with their supervisors.
- Employees want to know exactly what is required of them.
- Employees prefer contributing to setting their job-related duties than having someone else do it.
- Employees want to know about their future with the organization.
- They seek opportunities for personal growth and self-development.
- They want to contribute to the overall organization's objectives.
- They are motivated to achieve and accomplish and to be recognized for their accomplishments.

Finally, what are the techniques used in obtaining higher productivity? H. Z. Levine, in a recent article on the subject, listed the following methods:[46]

- Goal setting
- Compensation/rewards
- Team development
- Behavior strategies
- Management development
- Job retraining
- Feedback
- Quality of work-life programs

[46] H. Z. Levine, "Efforts to Improve Productivity," *Personnel,* **60** (January–February 1983):4–10.

Certainly the list could continue. The point, however, should now be clear to the manager. Supervisory personnel and employees have a dual role to fulfill in work performance and job satisfaction. It is a two-way street. Both parties must first realize that fact and motivation should follow—not just for the employees, but just as importantly, for the manager.

FREEDOM

Freedom within an organization would seem to imply that a chaotic environment exists. After all, if employees enjoy freedom who will do their work? A better question to ask is—the employee is free for what?

Abram Collier suggests that employees desire the freedom of thought and action, and freedom from worry.[47] He believes these are important values because they support and make possible the creative society of humans. As a manager it is necessary to consider these freedoms when dealing with your employees.

Where freedom of thought exists there will be differences of opinions. Out of these differences will arise conflicts; these conflicts should be viewed as necessary for creative work. The supervisor should not work to eliminate differences but encourage an atmosphere of understanding and respecting fellow employees.

Freedom of action refers to the employees' desire to find new and better ways of looking at job-related problems. A person wants to exercise control over his job. When this is taken away, he will become passive and apathetic. When given the opportunity to control his own job, the employee will be more likely to make a personal commitment to produce results. The supervisor can motivate employees by holding them accountable for responsibilities for tasks within their job scope. Since people know their own limits and capabilities, those who actively participate in this type of freedom will produce results.

There is also the freedom to do the minimum amount of work to avoid being fired or penalized. Yankelovich and Immerwahr call this "discretionary effort" and believe that new jobs and technology are creating a high-discretion workplace.[48] "Discretionary effort" is the difference between the maximum effort that could be used in doing one's job and the minimum effort to get by.

---

[47] Abram T. Collier, "Business Leadership and a Creative Society," *Harvard Business Review on Human Relations* (New York: Harper & Row, 1979), p. 420.

[48] Daniel Yankelovich and John Immerwahr, "Let's Put the Work Ethic to Work," *Industry Week*, September 5, 1983, p. 33.

It is the supervisor's job to develop a work environment that allows each employee the freedom to grow. Tasks must be consistent with employee needs. Since each employee has different needs, the supervisor should evaluate them and plan them into the work whenever possible. If you have an employee who is motivated by an affiliation need, this could be met by giving him or her the opportunity to work with others.

As outlined earlier, Fredrich Herzberg believed the things that motivate workers to work harder are (1) achievement, (2) recognition, (3) the work they perform, (4) responsibility, (5) opportunity for advancement, and (6) growth.[49]

Creative Freedom

When managers are asked how they manage creativity, the overwhelming response is something like, "not very well." Creative freedom by its nature is both frightening and exciting to the supervisor whose job it is to manage it. Abram Collier discovered that creative people are both egotistic and self-sacrificing—driven by an urge to create and grow.[50] He further thought if man is allowed to be free, society's fear is that he will become creative. From this creativity will develop growth and trouble.

The manager must develop an environment in which creativity can flourish; continued business success depends upon it. Even as society is advanced by creativity, so a business advances through its creative efforts.

The supervisor needs to challenge his employees to approach their jobs creatively. Creative people, however, will not be ready to agree with each other. Disagreement of this type can be good for the organization since it helps the manager to make wise decisions based upon different opinions.

The by-products of creative effort are fame, power, freedom, and happiness. The creative employee is trying to satisfy these needs and should be viewed as an asset to his department.

Individualism

Humans are different, and no matter what the corporate mold may dictate, the supervisor must deal with these differences. Individualism does not mean that your employees will show up for work in a suit and tennis shoes, nor are they likely to conduct themselves in a manner that would

---

[49] Kenneth H. Killen, "Motivation to Work," *Midwest Purchasing*, September 1983, p. 9.
[50] Abram T. Collier, "Business Leadership and a Creative Society," *Harvard Business Review on Human Relations* (New York: Harper & Row, 1979), p. 422.

be considered socially unacceptable. Individualism recognizes that each person will approach a task differently with his own unique solution. An individualist values his self-expression and seeks to work to his potential. The corporation has well-defined role behaviors and expectations, but amidst it all the individualist will seek self-identity. The old saying, "treat others like you want to be treated," will not work with these people—they are different and expect to be treated differently than you want to be treated yourself.

How should the supervisor manage and motivate the individualist? Realizing that each person has special talents, the supervisor needs to match work assignments to each employee. It is quite likely an employee may be suited for one assignment but unsuited for another. Work should be challenging and worthwhile, giving the employee a chance for personal growth.

Individualist's like to think of themselves as individuals, not members of the corporate team. They like to find their own method of solving a problem. Because of this, motivation must be done on an individual basis. Recognition is probably the most important motivator to this type of person.

Individualism is compromised when the individual joins a team. Initiation into the team is done to reduce individualism and build a team culture. For example, in one computer company there was a "rite of initiation." This rite of initiation, through which every member of the team passed, was "signing up," when the member agreed to do whatever was necessary to make the project successful. The heart of "signing up" entailed

- Agreeing to forsake family, hobbies, and friends
- Volunteering to work, giving it your heart and soul
- Agreeing to do something that had not been done before
- Associating with a highly intelligent, motivated, fast-pacing peer group
- Total commitment to the project[51]

Employee Rights

Employee rights are not to be thought of as employee freedom—there is a difference. Freedom is given to an employee by the corporation. Rights are something that cannot be taken away. Some rights are free speech,

[51] Paraphrased from Tracy Kidder, *The Soul of a New Machine* (Boston: Little, Brown, 1981), pp. 63–66.

privacy, due process, and dissent. These terms conjure up images of division, clashes, and exploitations which have no place in the workplace. Even though these rights are guaranteed by the Constitution, once an employee enters the workplace, for all practical purposes, he becomes rightless.

An employee of American Telephone and Telegraph[52] was fired after he was arrested at a political rally. The employee felt he had been fired without cause because of his political beliefs and sued. The court ruled the firing was justifiable since public policy does not prohibit an employer from firing an employee at will.

Let us examine a few of these rights. The first is the right to privacy. Employees want to control what facts, relevant or irrelevant, are released to others. They do not want any facts released without their consent. This would also include work-related information and the right to maintain their confidentiality.

The right of due process recognizes the employees' right to have a procedure available to them with the organization to discuss any infringement of their rights.

There is a new attitude developing called "psychology of entitlement."[53] This attitude teaches that each person deserves certain rights without earning them. They are entitled to them for being members of society. Society owes people a high income and a good job. Both blue-collar and white-collar workers want changes in employee rights and how they affect their jobs. Whether this is a passing fad or a new dimension in future employee relations remains to be seen.

Need for Order

There is a new demand being placed upon corporations today to demonstrate a concern for the well-being of their employees. Employees, both white-collar and blue-collar, want to be recognized as individuals and given freedom to perform their jobs to the best of their ability. The company benefits from creative freedom, but there exists a need for structure and order also.

Currently industry is using participatory management and job enlargement programs to address these issues and motivate the new breed of employee. The following chart can be used by the supervisor to evaluate

[52] Joseph A. Fernandez, "Employee Rights: Radical Propaganda or Reality?" *ABA Banking Journal,* September 5, 1983, p. 152.
[53] Erwin S. Stanton, "A Critical Re-evaluation of Motivation, Management, and Productivity," *Personnel Journal,* March 1983, p. 212.

individual employees and determine what leadership style is required to an individual basis. This chart was developed by Erwin S. Stanton.[54]

FACTORS SUGGESTING A MORE DIRECTIVE MANAGEMENT STYLE	FACTORS SUGGESTING A MORE PARTICIPATIVE MANAGEMENT STYLE
The employee tends to be more leisure-oriented, rather than seeking fulfillment through work	The employee seeks to fulfill many of his or her ego and psychological needs through expression at work.
The employee's job experience is such that he or she lacks the requisite qualifications to take on greater responsibilities.	The employee has the necessary intelligence, education, and experience to take on additional responsibilities.
The employee's educational or skill level is relatively modest.	The employee is interested in having more of a say in matters that affect him or her on the job, and wants to participate in decision making with management.
The employee has a personal reluctance to take on additional job responsibilities.	The employee does not feel anxious, uncomfortable, or insecure when faced with relatively unstructured and ill-defined work situations.
The employee requires a relatively structured, clearly defined, and essentially nonambiguous work environment in order to perform most effectively.	The employee is sufficiently self-reliant and self-confident to not need close and supportive supervision from his or her superior.
The employee needs fairly close and supportive supervision.	The employee identifies with the goals and objectives of the organization.
The employee does not express a personal interest in becoming involved in decision-making activities.	
The employee fails to identify sufficiently with the goals and objectives of the organization.	

Job Enlargement

Job enlargement is easier to accomplish then participative management. The theory is that if the variety of operations is increased workers will be motivated from job satisfaction. It is the supervisor's job to match the assignment with the employees. In reality this is very difficult to do. Individuals differ in what they want and expect from their job. The job that Sam finds boring is acceptable to Harry. The question then is, to whom do you tailor the job? Do you develop a different sequence for each person in the department?

[54] Erwin S. Stanton, "A Critical Re-evaluation of Motivation, Management, and Productivity," *Personnel Journal,* March 1983, p. 212.

MORALE

People in every organization are working together within a subtle environment of attitudes toward supervisors, attitudes toward physical conditions, attitudes toward financial rewards, attitudes toward products, and the like. Since attitudes affect the way groups and individuals behave in their work responsibilities, an understanding of work-related employee attitudes is centrally important to personnel administration. The manager's responsibility is to ensure that the conditions are right to maintain high employee morale. The employee relations department plays an active role in helping managers build and maintain an effective degree of employee morale throughout the company. A confidential study of a nationwide manufacturing company showed that the plant with the highest production rate had the highest morale, while the plant with the next highest production rate had the lowest morale.[55] Four aspects of employee morale will be discussed in this section: (1) the nature of morale, (2) relationships of morale to employee behavior, (3) method of measuring morale, and (4) Building morale through managerial effort.[56]

The Nature of Employee Morale

Employee morale is a concept that describes the level of favorable or unfavorable attitudes of the employees collectively toward all aspects of their work—the job, the company, their tasks, working conditions, fellow workers, supervisors, and so on. The way individuals think and feel contributes a lot toward their attitudes. Happiness cannot be equated with higher job satisfaction or morale. Happiness of a person is affected by the environment the person is surrounded by. The company or its managers cannot alone fulfill the needs to achieve happiness in a person. It is a question that goes beyond the responsibility of the company or its managers. Satisfaction as a requirement of an employer-employee relationship stops short of this happiness goal. Neither is morale described as the absence of conflict or the presence of good personal and interpersonal adjustment, nor is it group cohesion, loyalty, or acquiescence to the group's demands or superior involvement in company or group goals. Morale is the extent to which an individual's needs are satisfied and the extent to which the individual perceives that satisfaction as stemming from his total job situation.

[55] Sayles and Strauss, *Human Behavior in Organizations* (Englewood Cliffs, N.J.: Prentice-Hall, 1966), p. 144.
[56] McFarland, *Personnel Management* (New York: Macmillan, 1968), p. 483.

Morale is present in organizations in varying degrees. Morale is an attribute of the individual, but it is an attribute which exists primarily with reference to a group of which he is a member. Morale is a function of the amount of personal need satisfaction the individual can get from the activities and aspirations of a group with which he identifies. Morale fluctuates over time—sometimes high, sometimes low. Morale has a contagious quality in the sense that in the interaction and communication processes in an organization, employees transmit their changing attitudes and opinions to each other. A downturn in morale moves faster than an upturn, and it is hard to build low morale into high morale and easier to experience a lowering of morale when dissatisfying factors or events appear.

Attitude plays an important part in determining employee morale. An attitude is the way an individual tends to interpret, understand, or define a situation or his relationships with others. Attitudes are the individual's likes or dislikes directed toward persons, things, or situations or combinations of all three. Attitudes become deep-seated attributes of the individual's makeup. They can change, but tend to change slowly. Individuals tend to hold their attitudes firmly, resisting the forces of attitude change. This firm behavioral quality of an individual has a great impact on generating favorable or unfavorable morale for himself and influencing co-workers. When the individual's attitudes change materially from those of the group, he is likely to drift away from it, or to be expelled, or cut off by the group from some or all of its benefits.[57]

Morale and Employee Behavior

Psychologists have concluded that attitudes alone are not an adequate basis of predicting behavior, since attitudes are only one of several components that determine behavior and action.[58] Managers have taken an interest in problems relating to morale, attitudes, and job satisfaction not only because of their potential implications for employee performance and productivity, but for other aspects of the employer-employee relationship as well.

Studies on turnover show that persons with positive job attitudes are consistently found to be less likely to resign than those with negative attitudes. The relationship between attitudes and absences appears to be smaller in magnitude but similar in direction—people with positive job attitudes appear to be absent less frequently.[59] Views held by the human

---

[57] Victor H. Vroom, "Employee Attitudes," *The Frontiers of Management Psychology* (New York: Harper & Row, 1964), pp. 147–143.

[58] Ibid., p. 135.

[59] Ibid., pp. 136–137.

relations experts in industry have frequently suggested that the higher the morale, the higher the productivity, and conversely, the lower morale, the lower the productivity. Two studies at the University of Michigan are notable. Katz, Maccoby, and Morse applied four indices of morale in studying an insurance firm: pride in one's work group, intrinsic job satisfaction, company involvement, and financial and job satisfaction. Only pride in one's work group showed a distinct relationship to productivity.[60] The research found a lack of significant relationship between employee attitudes and productivity.

Although there is no direct relationship between productivity and morale, the need to have high morale and the importance of morale cannot be ignored. Morale remains a valid challenge for the leadership skills of managers, since productivity is not the only value organizations are striving for. Sustained low morale may eventually lead to manpower costs the company will pay, such as high turnover, absenteeism, retraining, and selection costs. Continued low morale will eventually lower productivity. Management cannot ignore the trend of decline of morale on the ground that it will not make any difference in productivity. On the other hand, high morale appears to be a predisposing factor to effective employee behavior and it is therefore worth considering as an organizational and managerial objective.

Morale is also important in group dynamics. Kurt Kewin, a professor of social psychology at the University of Iowa and the founder of group dynamics, observed through his experiements in industry that leadership attitudes had a direct correlation with worker morale and productivity.[61]

One study that is often cited regarding dehumanization of division of labor was conducted by C. H. Walker and R. H. Guest. They found that many workers disliked many things associated with their specialized jobs. Things such as mechanical pacing, repetitiveness of operation, and lack of a sense of accomplishment led to low morale. A relationship existed between the number of operations performed and the overall interest that the employee had in his job.

There are four relatively independent areas that may either contribute to or detract from job satisfaction. They are intrinsic job satisfaction, satisfaction with the company, satisfaction with supervision, and satisfaction with rewards and the opportunity for upward mobility.[62] A number of companies are using job enrichment to produce excellent results in the

---

[60] Daniel Katz, Nathan Maccoby, and Nancy Morse, *Productivity, Supervision, and Morale in Office Situation* (Ann Arbor: University of Michigan Survey Research Ctr., 1950).

[61] Robert M. Fulmer, *The New Management* (New York: Macmillan, 1974), p. 134.

[62] Robert L. Kahn, "Productivity and Job Satisfaction," *Personnel Psychology,* Autumn 1960, pp. 275–287.

improvement of worker morale. It is not feasible with all jobs. It can be mismanaged, and some people do not want it—but overall it offers a strong probability of favorable results.

As a motivator of workers, morale has far outdistanced money. "If you want to motivate the worker, don't put in another water fountain; give him a bigger share in the job itself."[63]

Measuring Morale

Morale is measured through surveys commonly known as morale surveys, job-satisfaction surveys, and attitude surveys.[64] They consist of a set of questions designed to obtain employees' opinions on various aspects of their employment. Morale can be high or low. It is very difficult to really define the level of morale if there are no accurate benchmarks. However, through the regular use of attitude surveys, one can observe changes from one period to another. Many companies express a desire to learn about areas of minor concerns that require attention by management before they become major uncontrollable situations. Through attitude surveys, management can determine their strengths or weaknesses in matters concerning employee morale and at the same time locate feelings and sources of irritation which require remedial action.

When conducting attitude surveys, the employee relations department faces the problems of introducing and applying new ideas to launch such survey programs. Planning, technical, and administrative issues arise.

During the planning stage the management has to realize that there is a "people problem" that requires attention by management, and a project is required to correct the situation which may not result in productivity gains but in other benefits. Someone should be assigned for the survey. The question arises whether to use consultants from outside or someone within the company to conduct and interpret the attitude surveys. If someone from within the company is chosen who has the proper qualifications, then a minimal use of outside assistance is needed. The qualifications of such a person should include familiarity with techniques of attitude surveys, a knowledge of statistical tools, and the ability to earn and retain the confidence of many persons at all levels in the company.

Technical problems include the design of the attitude surveys, the selection of sample size, methods of administering the survey itself, and the analysis of the results. Another decision centers on where and when to give the survey.

---

[63] Robert M. Fulmer, *The New Management* (New York: Macmillan, 1964), p. 360.
[64] William B. Wolf, *The Management of Personnel* (Belmont, Calif.: Wadsworth), p. 328.

Some of the methods of attitude surveys are as follows:

- Printed questionnaires which measure primarily qualitative opinions. These are questionnaries which include yes/no, true/false, and checklist-type questions. They also provide areas for written-in comments.
- Interviewing techniques—interviews are both the guided (directive) and unguided (nondirective) type.[65] This is again for qualitative information.
- Attitude scales which provide an objective and quantifiable basis for measurement.[66] Surveys must fulfill the statistical requirements of validity and reliability, the same concepts as those required for psychological tests. Brayfield and Roth have stated a set of criteria for developing an attitude scale.[67]
- Provide an index to "overall" job satisfaction rather than to specific aspects of the job situation.
- Applicable to a wide variety of jobs.
- Sensitive to variations in attitudes.
- Provide reliable and valid measurement.
- Brief and easily scored.
- Should be simple, interesting, and realistic—and must evoke cooperation from both management and employees.

The remaining technical difficulties in administrative tasks is the tabulation and analysis of the data. Desired tables must be constructed during the design of the questionnaire so that the completed questionnaires are returned and the desired data can be filled in. Analysis is the task of the survey researcher. Analysis will not produce findings or results beyond limits built into the original survey.

After the completion of the survey, the presentation of the findings to key members of management and the decision affecting the company and the employees must be made. These elements are important:

- Communicating results to the people involved
- Taking actions to correct the problems discovered by the survey
- Feedback of information, both good and bad, to those who participated in the survey and to those in a position to do something about the problems it uncovers

---

[65] Morris S. Vitelas, *Motivation and Morale In Industry* (New York: Norton, 1953), p. 224.
[66] Falton E. McFarland, *Personnel Management* (New York: Macmillan, 1968), p. 493.
[67] Ibid., p. 493.

Building Company Morale

One of the best ways to build company morale is to have a successful, soundly managed company, with top management interested in the employees at all levels. Practice occurs in the present and not in the future. If, therefore, management has to control future events, particularly events involving the interactions of persons, they must do something here and now which will have the desired consequences.[68]

Four widely used areas of morale building are[69]

1. Management of attitudes and attitude surveys
2. Problems of organization design
3. Employee participation
4. Handling problems of conflict

The management of attitudes (the daily changes and maintenance of positive attitudes among the members of management and workers) is very important. To cultivate favorable attitudes with morale-building objectives is one of the most important characteristics of the managerial process.

The problem of morale in an organization breaks down into basically two parts: (1) the daily problems of maintaining internal equilibrium within the organization (maintaining a kind of social organization in which individuals and groups through working together can obtain human satisfactions that will make them willing to contribute their services to the economic objectives of cooperation), and (2) the daily problems of diagnosing possible sources of interference, of locating sore spots, and of reducing human tensions and strains among individuals and groups.

Maintaining internal equilibrium in an organization within the social organization of the plant involves keeping the channels of communication free and clear so that orders are transmitted downward without distortion and relevant information regarding situations at the work level are transmitted upward without distortion to those levels at which it can be put to best use. This involves getting the bottom of the organization to understand the economic objectives of the top, and the top management to understand the feelings and sentiments of the bottom.

The readiness of an employee to accept a change depends on personal

[68] F. J. Roethlisberger, *Management and Morale* (Cambridge, Mass.: Harvard University Press, 1975), p. 177.
[69] Dalton E. McFarland, *Personnel Management* (New York: Macmillan, 1968), p. 496.

and organizational factors. The greater the participation on the part of the people subject to a proposed change, the greater will be its acceptance.

The use of engineering teams requires a change in organizational roles and in managerial style, particularly at the supervisor level. The success of teams in the engineering organization depends on how willing such supervisors are to make teams work and to change their supervisory style accordingly. Senior management needs to be committed and supportive of first-level supervisory efforts to change their attitudes and behavior.

Supervisory-level attitude and behavior changes to support team management include

- Egalitarian values
- Willingness to meet with team members on a peer basis
- Reciprocity in dealing with team members
- Disposition for participative management
- Decisions by the most knowledgeable persons
- Management by merit, explanation, and the consent of the managed
- Trust, confidence, commitment, and loyalty to others
- Willingness to work diligently at communication, particularly in listening more than ever before
- Willingness to delegate authority and responsibility to the team members
- Sharing of information, decisions, results, and rewards

These attitudes and behavior do not suddenly appear; their development and acceptance takes time. This time may be lengthened by the supervisor's sense of loss of traditional prestige and power attached to the traditional supervisory role. Supervisors may not want to give up their prestige or the power to "control" their workers. Then too, they may not really understand how to function under the more participative environment. All too often a traditional supervisor perceives workers as being indifferent to team performance as part of an overall organizational strategy. If the organization is unionized, supervisors may feel that employee participation will be used as a means to further the union's objectives rather than the team's purposes. Some organizations have reduced this possibility by taking prompt action to establish worker-manager teams to improve organizational performance.

SUMMARY

In this chapter we discussed the management function of motivation—the artful style of dealing with people in such a manner that it brings out their best attitude and performance. People are motivated by diverse factors of their background, organizational ambience, work experience, interpersonal relationships, home life, and the leadership style they face. The chapter summarized motivation concepts of theorists in the field. In addition, we drew on the experiences of practitioners in the field. We provide general guidance for working with people, managing creativity, dealing with the problem employee, and using reward systems to motivate people. Work performance motivational factors, productivity, job satisfaction, individualism, job enlargement, and morale are topics which finished out the chapter.

Planning and organizing are covered in Chapter 5.

BIBLIOGRAPHY

Argyris, Chris, "Personality vs. Organization." In Keith Davis and John W. Newstrom, *Organizational Behavior: Readings and Exercises,* 6th ed. New York: McGraw-Hill, 1981.

Beck, Robert C. *Motivation Theories and Principles*. Englewood Cliffs, N.J.: Prentice-Hall, 1978.

Blake, Robert R., and Jane S. Mouton. "Increasing Productivity Through Behavioral Science." *Personnel,* May–June 1978.

Bushardt, Stephen C., Roberto Toso, and M. E. Schnake. "Can Money Motivate." *Management World,* November 1981.

Collier, Abram T. "Business Leadership and a Creative Society." *Harvard Business Review on Human Relations*. New York: Harper & Row, 1979.

"The Creative Temperament." *Management Review,* vol. 71, July 1982.

Davidson, Richard. "Motivating the Underachievers." *Supervisory Management,* January 1983.

Davis, Keith. *Human Relations at Work.* New York: McGraw-Hill, 1967.

Davis, Keith. "Low Productivity? Try Improving the Social Environment." In Keith Davis and John W. Newstrom, *Organizational Behavior: Readings and Exercises,* 6th ed. New York: McGraw-Hill, 1981.

Ouchi, William G., *Theory Z: How American Business can Meet the Japanese Challenge.* Reading, Mass.: Addison-Wesley, 1981.

Fernandez, Joseph A. "Employee Rights: Radical Propaganda or Reality?" *ABA Banking Journal,* September 5, 1983.

Fulmer, Robert M. *The New Management.* New York: Macmillan, 1974.

Gellerman, Saul W. "The Meaning of Motivation: The Self-Concept." In Keith Davis and John W. Newstrom, *Organizational Behavior: Readings and Exercises,* 6th ed. New York: McGraw-Hill, 1981.

George, Claude S., Jr. *Management for Business and Industry.* Englewood Cliffs, N.J.: Prentice-Hall, 1970.

Hackman, J. R., and J. L. Suttle. *Improving Life at Work: Behavioral Science Approaches to Organizational Change.* Glenview, Ill.: Goodyear, 1977.

Herzberg, Frederick. *The Managerial Choice.* Irwin, Ill.: Dow Jones, 1976.

Hicks, Herbert G. *The Management of Organizations: A Systems and Human Resources Approach,* 2nd ed. New York: McGraw-Hill, 1972.

Hicks, Herbert G., and W. Jack Duncan. *The Management of Organizations*. New York: McGraw-Hill, 1972.

Kahn, Robert L. "Productivity and Job Satisfaction." *Personnel Psychology*, Autumn 1960.

Katz, Daniel, Nathan Maccoby, and Nancy Morse. *Productivity, Supervision, and Morale in Office Situation*. Ann Arbor: University of Michigan Survey Research Center, 1950.

Kidder, Tracy. *The Soul of a New Machine*. Boston: Little, Brown, 1981.

Killen, Kenneth H. "Motivation to Work." *Midwest Purchasing*, September 1983.

Kleiner, Brian H. "Integrating Major Motivational Theories." *Journal of Systems Management*, February 1983.

Levine, H. Z. "Efforts to Improve Productivity." *Personnel*, vol. 60, January–February 1983.

Likert, Rensis. "The Principle of Supportive Relationships." In Keith Davis and John W. Newstrom, *Organizational Behavior: Readings and Exercises*, 6th ed. New York: McGraw-Hill, 1981.

Lorsch, Jay W. "Situational Theories of Behavior." In Keith Davis and John W. Newstrom, *Organizational Behavior: Readings and Exercises*, 6th ed. New York: McGraw-Hill, 1981.

McClelland. "Characteristics of Achievers." In Keith Davis and John W. Newstrom, *Organizational Behavior: Readings and Exercises*, 6th ed. New York: McGraw-Hill, 1981.

McFarland, D. E. *Personnel Management*. New York: Macmillan, 1968.

Peters, R. S. *Concept of Motivation*. New York: Humanities Press, 1960.

Phillips, Jack J. "Rewarding Employees Effectively: Substance and Style." *Supervision*, November 1982.

Quick, Thomas L. *The Quick Motivation Method*. New York: St. Martin's Press, 1980.

Roethlisberger, F. J. *Management and Morale*. Cambridge, Mass.: Harvard University Press, 1975.

Rosembaum, Bernard L. *How to Motivate Today's Workers: Motivational Models for Managers and Supervisors*. New York: McGraw-Hill, 1982.

Sanzotta, Donal. *Motivational Theories and Applications for Managers*. AMACOM, American Management Association, 1977.

Sayles and Strauss. *Human Behavior in Organizations*. Englewood Cliffs, N.J.: Prentice-Hall, 1966.

Spitzer, Dean R. "Thirty Ways to Motivate Employees to Perform Better." *Training*, March 1980.

Stanton, Erwin S. "A Critical Re-evaluation of Motivation, Management, and Productivity." *Personnel Journal*, March 1983.

Terpstra, David E. "Theories of Motivation—Borrowing the Best." *Personal Journal*, June 1979.

Vitelas, Morris S. *Motivation and Morale In Industry*. New York: Norton, 1953.

Vroom, Victor H. "Employee Attitudes." In *The Frontiers of Management Psychology*. New York: Harper & Row, 1964.

Yankelovich, Daniel, and John Immerwahr. "Let's Put the Work Ethic to Work." *Industry Week*, September 5, 1983.

5
PLANNING AND ORGANIZING

In this chapter we will discuss the planning and organizing responsibilities of a team manager. This chapter, aimed at both team leaders and leaders-to-be, attempts to convince them that there is nothing quite so important as the team's future. When planning is understood and undertaken by those responsible for a team, then the team can follow the evolution of a central plan of action through constantly changing circumstances in the future.

The second purpose of this chapter is to describe how a team leader can get the team organized. Within organizations such as a team there are leaders and followers who work together to accomplish results—the objectives and goals of the team. Team members cooperate in doing what has to be done to reach the desired end result. Cooperation requires that the team members discuss, negotiate, and reach agreement on the individual and collective roles in the team's operation.

WHAT IS PLANNING?

Planning deals with the creation of something that does not currently exist. A team's planning involves the organization's future direction. In deciding where the team is going in that future, the team members must consider several key questions:

- What have we done in the *past*?
- What are we *presently doing*?
- What should we do in the *future*?
- What should we *not be doing* in the future?
- How are we going to get *where we want to go*?

The end result of team planning is a prescription of where the team is going and how resources will be used to get there. Team plans reflect decisions that have been made on the purpose of the team. But a team's success does not rest solely on the design of its plans. Such plans have to be effectively implemented. Thus, for a team to get where it wants to go in the future, two essential steps are required:

1. Team planning
2. Implementation of the team's plan

Team planning involves the development of certain key planning elements:

- Team objectives
- Team goals
- Team strategies
- Team organization structure
- Team member roles
- Team "systems" support
- Team style
- Team resources

Implementation of the team's plan entails the following

- Development and use of an effective *leader* and *follower* style
- Productive use of human and nonhuman resources
- Design, development, and implementation of information systems to track and report team progress
- Design and implementation of a control system to monitor team progress and when necessary reallocate team resources to more effectively accomplish team objectives and goals
- Replanning as necessary

Team planning and implementation will be discussed in the following sections.

Team Objectives

A good place to start planning for a team is to ask the question, What "business" is the team going to follow? Consideration of this question raises the issue of the team's objective. A team's objective is the future position or destination that the team members wish to reach in its "business." The accomplishment of the team's objective contributes directly to the mission of the parent organization—a contribution that should be measurable. Some examples of the team objectives include the following:

- *Product design team:* Facilitate the design of a quality product that can be manufactured and serviced at a minimum cost.
- *Project engineering team:* Direct and integrate all technical aspects of the design and development process of product X.

- *Project management team:* Insure the accomplishment of the project on or ahead of schedule, at or below budgeted cost, and at the required performance level.
- *Production team:* Assemble product X on a timely and cost-effective basis.
- *Quality Circle Team:* Participate with management in improving work and the work environment in the organization.
- *Marketing task force team:* Determine if there is a viable market for product X under development in the company.

A team's objective may be stated in quantitative or qualitative terms—or some combination thereof. For example, a large equipment manufacturer appointed a task force team to "install and operate an automated flexible manufacturing system capable of reaching and maintaining a production cost standard in our X product line." An electronics firm appointed and advanced development team "to develop a strategy to leapfrog" the existing state-of-the-art semiconductors by X factor.

The proper selection of a team's objective is the first essential step in the management of the team's work. Composition of the team's membership is dependent on that objective. Team objectives are supported by team goals.

Team Goals

The distinctive features of team goals are their specificity and time-based measurement points. Attainment of a goal signifies that if present utilization of resources continue, the achievement of the team's objective will be attained. For instance, in the management of a construction project, the completion of a work package in the project work breakdown structure means that progress has been made toward the objective of delivering the project on time, within budget, and in satisfaction of its operational objectives. A company might set up schedule goals for reaching an objective of 15% return on a piece of installed production equipment, or select increments of productivity gains to be attained over a period of time.

Goals play an important part in the planning for teamwork. Individuals can be identified as responsible for certain goals and then held responsible for accomplishing those goals.

Team Strategies

A strategy is a series of prescriptions designating courses of action on how resources will be used to support team purposes. In the larger organizational context, strategies are the key prescriptions for the positioning of the organization in its competitive and environmental systems. The existence of strategies presumes that these environmental and competitive threats and opportunities have been identified, costs and risks have been assessed, and organizational strengths and weaknesses have been evaluated. Strategies provide the parallel pathways that the organization must follow to get where it wants to go as well as mapping the course to provide a tactical chart of how the resources of an enterprise are to be mobilized to accomplish mission, objectives, and goals. When a team is organized, an important step has been taken to mobilize resources to accomplish a specific purpose.

In addition to establishing the general direction for committing resources, strategies prescribe team organizational design, key roles, desired key cultural characteristics, and organizational systems to support team activity. When strategies are finalized, it means that the strategic planning process has been carried out in the organization. What remains is the use of resources through administrative and cultural means: structure, roles, systems, and style to execute strategies.

A couple of examples will illustrate the use of strategy and how it affects organizational destiny. A large electrical/electronic system manufacturer states its strategy as "emphasizing high growth and high technology through a commitment to the factory of the future." Product development teams are used to carry out this strategy. The John Deere Company, a large manufacturer of construction and farm equipment, emphasized past key strategies in its 1982 Annual Report in the following manner:

> . . . decentralized . . . management that emphasizes delegating as much decision-making authority as possible to the person closest to the scene. [p. 3.]
>
> . . . diversify both in products and geographically. Plans to become a manufacturer of construction equipment and grounds care equipment were developed. The company also began establishing manufacturing facilities outside North America. The second decision was to cease production of tractors powered by two-cylinder engines, long treasured by farmers for their dependability. Though the two-cylinder tractors still were at their peak of popularity, they had reached the limit of power and it was recognized the company must change to meet the power requirements of the future. The changeover involved the largest single capital investment in the company's history. The decision proved to be correct. Three years after the new tractors were introduced the company became the leading producer of farm machinery in the free world. [p. 3.]

The early success of the Control Data Corporation, whose sales are in the $5 billion range, can be attributed to several key and enduring strategies formulated and followed by the founder and chairman, William C. Norris. These strategies are as follows:

- Concentration on large-scale computers with scientific and engineering applications
- No challenge to the IBM Corporation in its leadership in computers
- Ownership of Commercial Credit, a financial services company with a strong cash flow that helped control Data to finance its growth in leader computers.
- Manufacture of peripheral equipment to supply other computer manufacturers

When strategies are developed, the planning phase has identified organizational purposes and prescribed how resources will be used to reach those purposes. What follows is the implementation of those plans through the use of organization structure, roles, systems, and managerial/follower style.

Team Organization Structure

Early principles of organization set forth various organizing techniques. Line and staff concepts and the vertical, hierarchical chain-of-command beliefs provided basic points of departure from which to organize activities. *Centralization, decentralization, functional, departmental, product, process, and geographical* are the primary patterns and techniques used today for structuring the organization. Alternative ways of organizing, emphasizing a team focus, have been tried in recent years. For instance, in 1974, the General Motors Corporation, as part of its strategy for downsizing their automobiles, adopted the "project center" concept as a way of organizing engineering resources to redesign automobile components. Project centers, composed of design teams, work on the design of common items of equipment and problems common to all divisions, such as frames, electrical systems, shock absorbers, and steering gear. When common equipment was designed for the car models in the various GM divisions (Pontiac, Buick, Cadillac, Oldsmobile, and Chevrolet) manufacturing and assembly processes would be common, with a resulting saving in product cost.

At the time the "project center" strategy was announced by GM, the authors speculated with their students about the possibility of this strategy ultimately bringing about a "look-alike" result in GM autos. This would come about as more and more common items of equipment were designed, resulting in autos whose product differentiation would be perceived by customers as little more than a facade. This may be what has happened. *Fortune* magazine, in the February 21, 1983 issue, reported that sales of the X cars have been falling, perhaps largely as a result of the usual decline in any model's sales over time. But, according to *Fortune*, ". . . the main reason for their coming demise, in the view of most industry experts, is that GM now has too many models too much alike."[1]

Within a structure, the performance of individual and collective roles plays an important part in maintaining the organizational balance. All too often such roles are inadequately defined, resulting in jeopardizing an effective strategy for managing an organization. Where people work together as a team, role definition is essential. However, this role definition may be done in an informal manner. For example, one company, Gore Associates, uses a "lattice" organization rather than a well-defined pyramidal hierarchy. Lines of communication are direct person-to-person; other attributes of this organizational form include

- No fixed or assigned authority
- Sponsors, not bosses
- Natural leadership defined by followership
- Objectives set by those who must "make them happen"
- Tasks and functions organized through commitments
- A structure that evolves from interpersonal interactions, self-commitment to know responsibilities, natural leadership, and group-imposed discipline

Teams evolve within the lattice structure usually by that individual component in the discipline or activity of the team. An individual may serve on several teams. The "lattice" organization grew out of the use of task forces in the company.[2]

Team Member Roles

The definition of the structure of an organization portrays the major territories that are assigned to each manager. Within each territory—production, finance, marketing, and so forth in a business enterprise—specific roles require identification and negotiation, particularly in terms of the interfaces of individuals with peers, subordinates, and supervisors. For a

[1] "X Cars Exit," *Fortune*, February 21, 1983, p. 12.

[2] Talk by W. L. Gore, "The Lattice Organization—A Philosophy of Enterprise," undated, W. L. Gore & Associates, Inc., Newark, Delaware. See also, "Classless Capitalists," *Forbes*, May 9, 1983, pp. 122–124.

team to be effective, each member must understand how to work with other members. These role interrelationships come to focus through work packages—discrete elements of related work—and are held together by accepted authority, responsibility, and accountability of the managerial team. They are used to identify personal responsibility and control work flows in the organization.

Work packages are level-dependent, becoming increasingly more general at each higher level and increasingly more specific at each lower level. A president would be expected to have primary responsibility to "set corporate objectives"; a contract manager would be expected to take the lead in developing corporate maintenance agreements for a profit center manager to use in supporting equipment that has been delivered to a customer. An individual on a team is designated as having primary responsibility for each work package. Others who have collateral responsibility involving the work package are also designated. When these collective roles have been specified there remains "no place to hide in the organization." If the work package is not completed on time or does not meet the performance standards, someone can readily be identified and held responsible and accountable for that work. Responsibility, authority, and accountability—the triad of personal performance in organized life—are not left in doubt when individual and collective roles have been adequately defined. This definition can be carried out through the process of linear responsibility charting (LRC).[3]

Team "Systems" Support

Systems provide the means by which both the regularly interacting management and organizational parts are pulled together to make the organization an ongoing entity. For example, within a manufacturing company, the production, marketing, financial, and engineering systems are organic to the operation of the enterprise. These organizational systems are supported by planning, control, motivation, information, and decision support systems. The manager uses the latter systems in an artful and scientific manner as the medium through which the organic organizational systems (production, marketing, etc.) are managed. Team effort is supported by other effort within the organization. For example, a product design team would depend on such support from budgeting and accounting, manufacturing, engineering, product engineering, human resources, computer services, systems engineering, product design, production planning, etc. In the workings of a team, leader and follower style are important.

[3] See p.184

Team Style

Much of the culture of an organization is dependent on the style followed by the leaders and followers. Leaders tend to develop and propagate a distinctive manner of expression and acting which becomes characteristic of the organization. Followers tend to evaluate their leaders. The collective style of the followers and leaders permeates a team, resulting in the team developing distinctive culture.

Whatever the manager does—or does not do—will set the tone for the style practiced in the organization. That style is important to the effective functioning of the organization.

Team Resources

The quality and quantity of the human and nonhuman resources provide support to everything else in an organization's effort: style, systems, roles, structure, strategies, goals, objectives, and mission. Without adequate resources nothing works well. The quality of the human resources depends on the state of the art of the knowledge, skills, and attitudes of people in an organization. In today's rapidly changing world, managers and professionals are running a race between obsolescence and retirement. Many people—and organizations—are losing that race.

Team Planning

A team effort must be planned, no matter how small the team's project. The team's plan addresses such questions as

- What is to be done?
- Who will do the work?
- When and how will the work be done?
- What resources are available to support the work?

A team plan may not need to be elaborate. A team's work is usually part of a larger effort under way in the organization that has its own overall plan. For example, the essentials of a team's plan would probably include the following:[4]

- A summary of the team's purpose that states briefly what is to be done and the methods and techniques to be used. *It lists the deliverable end products in such a way that when they are produced they can be easily identified and compared with the plan.*

[4] Paraphrased from Bertram N. Abramson and Robert D. Kennedy, *Managing Small Projects*, TRW Systems Groups, TRW Inc., 1969.

- A list of tangible and discrete goals, identified in such a way that there can be no ambiguity about whether a goal has been achieved.
- A work breakdown structure (WBS) that is detailed enough to provide meaningful identification of all tasks associated with the team's efforts.
- A strategy outlining how resources will be used to accomplish team objectives and goals.
- An activity network that shows the sequence of the elements of the team effort and how they are related (which can be done in parallel, which can start only when another is finished, etc.).
- Separate budgets (if appropriate) and schedules for the elements of the team effort for which some individual can be held responsible.
- An "interface" plan that shows how the team's effort relates to the rest of the world, most particularly with the "customer."
- An indication of the team review process—*who reviews the team's work, when, and for what purpose?*
- A list of key project personnel and their team assignments in relation to the work breakdown structure.

The creation of a plan serves to document the team's future effort. The process of going through the planning effort is as valuable as the resulting plan. By planning together, the team members begin to understand the team's purpose and the other members' roles in fulfilling that purpose.

Planning is the orderly and logical exploration of developing a vision of what a team might be in the future. Planning is a continuous process of seeking an answer to the question, What are we aiming for and why? Planning decisions are those decisions made today whose impact will be felt in the future. Planning links formal analysis and intuitive judgment in selecting a course for the team's future.

An important part of team planning is that planning required to organize the people to work in the team and in the larger organizational context in which the team is found. We begin this discussion by exploring a typical organizing process. This will lead us to an examination of the concepts of authority, responsibility, and accountability, three key elements in all forms of organizations.

Organizational Structure

Organizational structure is the arrangement of the elements of the enterprise in their relationships to each other. *Organizing* is the *process* of arriving at an orderly arrangement of the interdependent parts of an organization. This process of organizing applied to the enterprise as an entity or a component thereof—such as a team—includes

- Identifying the mission, objectives, goals, and strategy of the organizational unit
- Determining the work that needs to be done
- Classifying and arranging the work into manageable parts
- Defining the individual and collective roles required to do the work through the use of LRC techniques
- Defining policy methods, techniques, and procedures, with associated documentation required to support organizational roles
- Preparing position descriptions
- Assigning individuals to the positions
- Developing the individuals

This process is portrayed in Figure 5-1.

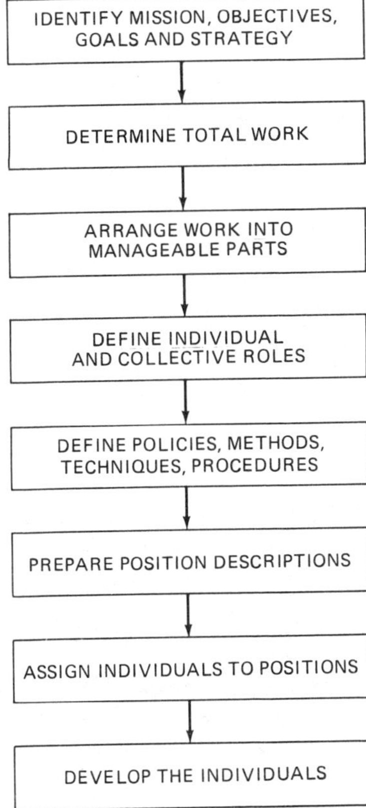

Figure 5-1. The organizing process.

ORGANIZATIONAL DESIGN

Organizational design is the conception and plan for the arrangement of elements that go into the organizational entity. Key concepts of authority, responsibility, and accountability are principal considerations in organizational design. The process of *delegation* permits the building of an organization.

Certain strategic conditions are a prerequisite to the satisfactory selection and use of a team form of organizational design. An awareness of these conditions and an understanding of their implications is essential to the successful conceptualization, design, development, and operation of a team management system.

First, the strategic planning and implementation philosophy used in the engineering enterprise must provide for the shared values in a culture which is supportive of team management techniques. The common denominator of such values requires an articulated and demonstrated willingness on the part of management and professional personnel to share organizational *resources, decisions, results,* and *rewards* at the organizational locus where team management will be used.

Second, the strategic mission, objectives, goals, and strategies developed for the engineering organization out of a strategic planning process should inherently contain the need for a team organizational design. On the surface this seems obvious; however, engineering organizations which provide routine engineering support would probably have limited opportunity to use project engineering teams. If the enterprise is in a high-technology industry, there would be abundant opportunity to use project engineering and management teams. Most, if not all, engineering organizations would find it useful to use some teams such as quality circles, management-professional teams, product teams and task forces.

Third, senior management must provide support, commitment, and a demonstrated behavior that sends a clear message to everyone: "Team management is important to the long-term management of this organization and we will do our best to make it successful."

First-level supervisors are a source of resistance to professional-manager teams; yet the support of these individuals is essential if meaningful success with team management is to take place. In a study of quality-of-work-life programs (QWL) Klein found that supervisors, while rarely showing open resistance, show concerns about job security and job definition. A third concern came from the amount of additional work generated by QWL programs that falls on supervisors.[5] Klein, by gathering evidence

[5] Janice A. Klein, "Why Supervisors Resist Employee Involvement," *Harvard Business Review*, September–October 1984, pp. 87–95.

Table 5-1. Supervisors as "Resisters"[a]

TYPE	WHY THEY RESIST	CLUES TO BEHAVIOR
Proponents of Theory X	The concept goes against their belief system.	Comments such as "Employees are children, not adults" and "Employees will just take advantage of the program to get out of work."
Status seekers	They fear losing prestige.	Unwillingness to let go of behavior associated with control. Fear of losing leadership role. Comments such as "Foremen can't be equal members of a team; they will always be the leader."
Skeptics	They doubt the sincerity and the support of upper management.	Comments such as "This program is no different from past ones. It will fade away in a few months"; "The problem is the next level up"; and "They don't really practice what they preach."
Equality seekers	They feel that they are being bypassed and left out of the program.	Comments such as "Why do we have to change before the employees do?" Non-support and "hands off" as problems arise.
Deal makers	The program interferes with one-on-one relationships with workers.	Comments such as "We've been stripped of our power" and "We have no control over the process."

[a] From Janice A. Klein, "Why Supervisors Resist Employee Involvement," *Harvard Business Review*, September–October 1984, p. 90.

at the plant level, concluded that each of five types or categories of supervisors has its own reasons for opposing employee involvement programs. These conclusions are contained in Table 5-1.

A coordinated strategy that would make supervisors an integral part of the change process toward QWL programs would include the following:

- Adequate training
- Allowing supervisors to participate in the design and implementation of QWL programs

- Maintaining supervisor responsibility and authority
- Encouraging supervisor-peer "networking"
- Recognizing that some competent supervisors will not adapt and will require replacement.[6]

Delegation

Delegation—the act of empowering one person to act for another—is essential to the design and development of any organizational unit. When a manager appoints a representative to act on the manager's behalf, an act of *delegation* has occurred. When the delegatee accepts and executes the power to act for another person, an act of delegation has been consummated. Delegation and redelegation create lines of authority, responsibility, and accountability between positions within an organizational structure. Redelegation in an organization is carried as far as necessary to ensure that the organization's work is assigned to a specific work package designed for performance by individual professionals in the organization. Delegation can occur in several different ways in modern organizations:

- From manager to manager
- From manager to professional
- From professional to professional
- From partner to partner through the "agency" power
- From organization to organization through the contracting process

The organization chart is a model of the delegation process. The lines on such a chart depict the flow of authority and responsibility conventionally shown as solid lines connecting organizational positions at each level in the chart. The lines of responsibility have a two-way characteristic: Authority for a work package is delegated to a delegatee by the delegator, but the delegator still retains responsibility and accountability for the delegatee's performance on the work package. The delegation process can also be carried out through several other means:

- Memoranda of agreement
- Acceptance of a "work order" which allocates funds, establishes schedules, and delineates a work package to be accomplished as part of a plan of work in the organization
- Acceptance of a budget

[6] Klein, op. cit., pp. 92–93.

- Acceptance of a plan of action to accomplish organizational mission objectives and goals
- Acceptance of a contract from a customer, supplier, union, etc.
- Through the agency power in a partnership by which one partner is authorized to commit the partnership in transacting business
- Agreement to do something through mutual agreement with a peer member of the organization

The best known way of delegation is through assigning work to a person occupying a position in the "chain of command" in the organization. However, other means also exist. Indeed, the act of delegation occurs when one person agrees to do something on behalf of another.

Virtually every organization has some kind of delegation process, however formal or informal. Most people recognize and know the responsibilities of someone who is "in charge," someone to whom they can go for those decisions which they are not empowered to make. The formal structure of an organization is an important model to guide in the making and execution of decisions in organizations.

A manager delegates work to others in the organization so as to be free of routine short-range decisions and work and have time to deal with

- Exceptions to the routine
- Policy issues
- Strategic planning
- Those decisions and actions which the manager has reserved for himself

In modern engineering organizations, more and more engineers feel that they report to more than one boss. Those who do are bothered by this, sensing that something is wrong. Yet in reality, they have encountered this two-boss enigma before:

- As a specialist supporting a project engineer and reporting to a functional supervisor
- As a child reporting to two parents
- As a primary and secondary school student, with responsibility to a home room teacher as well as to other teachers offering specialty courses
- As a member of a sports team and being held accountable by the coach, team captain, and members of the team
- As a member of a formal engineering organization and participating in the informal organization.

At any one time in the workings of a dynamic engineering organization its members can realistically ask the question, Who's really the boss? Or to use a humorous note—if the boss calls, get his name!

Authority

Authority is the legal right to command, direct, or exact obedience. A person who is invested with authority has the right to determine or judge what actions people are to take. Authority implies that power is held by a single person or redelegated to other persons who exercise the authority on behalf of the delegator. Authority exists in an organizational position or through a person becoming an expert in a field, or through influence or persuasion of others by the exercise of knowledge or experience.

Authority is essential to any group effort. The authority that is exercised by an individual comes from the organizational position occupied by the individual. Such authority is given through a grant or delegation from a higher authority level in the organization. The ultimate source of authority in organizations can be traced back to the right of private property in our society. In a business organization the shareholders elect the board of directors of a company. These directors have the authority given to them by the corporate charter and bylaws to manage the corporation on behalf of the shareholders. The authority of the board of directors is broad, is of a fiduciary nature, and is the starting point for the delegation and redelegation of authority within the organizational structure. An example of the authority and responsibility of the board of directors of one organization goes like this:

> The board of directors and its chairman has ultimate responsibility for the legal and ethical conduct of the company and its officers. It is the board's duty to protect and advance the interests of the stockholders, to foster a continuing concern for fairness in the company's relations with employees, and to fulfill all requirements of the law in regard to the board's stewardship. The board has an important role in counseling management on general business matters, as well as in reviewing and evaluating the performance of management. To assist in discharging these responsibilities the board has formed various committees to oversee the company's activities and programs in such areas as employee benefits, compensation, financial auditing and investment.[7]

[7] Taken from pamphlet: Hewlett Packard, Corporate Public Relations, 1501 Page Mill Road, Palo Alto, California 94304. Used by permission.

The residual authority that exists with the board of directors in a company is further delegated down the organizational structure or chain of command. For example, the board of directors of the Hewlett Packard Corporation has further delegated authority to the president of the corporation through the following delegation:

> The president and chief executive officer has operation responsibility for the overall performance and direction of the company, subject to the authority of the board of directors. Also, the president is directly responsible for the corporate development and planning functions and for HP [Hewlett Packard] laboratories.[8]

The delegation and redelegation of authority continues down through the chain of command until the individual worker is reached. The position description for the individual worker provides that individual with the legal right to carry out certain activities to perform a specified role in the organization.

When an individual joins an organization, the consummation of the employment contract creates an agreement whereby that employee agrees—at least implicitly—to accept the authority of a person in charge as a condition of the employment. By accepting the employment contract, the employee agrees to offer up certain portions of his physical and mental abilities—his private property—to aid in the achievement of the organization's objectives and goals.

The foregoing concepts of authority appear to be straightforward. To accomplish action in an organization one needs only to issue orders down through the chain of command. Those people who are in the chain of command will see that they are carried out. Unfortunately this is a very simple viewpoint of how the relationships of authority operate in modern organizations. It just does not work that way in the real world. The concept of legal authority, the right to act and demand obedience, is tempered by influence and power which modify the legal framework of authority. Most modern managers function through a combination of legal and informal authority in dealing with the persons that a manager must influence in order to accomplish the purposes of the organization. Simply being in a position of a manager does not mean that you can compel those who report to you to do certain things. Authority is like a coin: It has two sides. First, the *legal* authority that a manager has by virtue or occupying an organizational position to which authority has been delegated through

---

[8] Ibid.

the organizational structure. The other side is the *informal* authority that the manager has been able to develop with those persons who are the object of the authority. This two-sided nature of authority has been long recognized by management writers. For instance, Fayol defined authority as ". . . the right to give orders and the power to exact obedience."[9] However, he further noted:

> Distinction must be made between a manager's official authority deriving from office and personal authority, compounded of intelligence, experience, moral worth, ability to lead, past services, etc. . . . personal authority is the indispensable complement of official authority.[10]

Any view of authority must consider the two sides that are involved. This view of authority has implications for the team manager. One's ability to get things done depends on more than just the legal authority of a position. A team leader must develop informal and influential relationships with the persons with whom he works—peers, professionals, managers, and associates. Even the president of the United States, who has awesome legal authority in our society, recognizes the degree to which formal authority is limited by the ability to influence the Congress, the Supreme Court, and the citizenry of the land. All presidents have used informal contacts—often called the political process—to influence and supplement their formal, legal authority.

Power

Power is a concept frequently associated with authority. Power is the ability to unilaterally determine the behavior of others, regardless of the basis of that ability.[11] The exercise of power by a manager or a professional depends on the authority that is inherent in the organizational position that the individual holds as well as the manner in which the individual discharges the responsibilities of that position. Power arises within the structure and is reinforced through plans, policies, procedures, and prescribed techniques for the operation of the organization. A newly appointed team manager may be uncomfortable about the use of power in doing the management job. But the alert manager will find out rather

[9] Henri Fayol, *General and Industrial Management* (London: Sir Isaac Pitman & Sons, Ltd., 1949), p. 21

[10] Ibid.

[11] For example, see James D. Thompson, "Authority and Power in 'Identical Organizations,' " *The American Journal of Sociology*, **60** (November 1956).

quickly that many situations require a use of influence beyond what is inherent to the organizational position. If the newly appointed team manager feels that a superior-subordinate relationship is a prerequisite to the exercise of influence in the organizations, that manager is missing a fundamental point about power in modern organizations. The sources and uses of power spring from many relationships in the organization. Drucker has provided insight into the workings of power in organizations:

> The hierarchy does not, as the critics allege, make the superior more powerful. On the contrary, the first effect of hierarchical organization is the protection of the subordinate against arbitrary authority from above. A scalar or hierarchical organization does this by defining carefully the sphere within which the subordinate has the authority, the sphere within which the superior cannot interfere. It protects the subordinate by making it possible for him to say, "This is *my* assigned job." Protection of the subordinate underlies also the scalar principle's insistence that a man have only one superior. Otherwise the subordinate is likely to find himself caught between conflicting demands, conflicting commands, and conflicts of interest as well as of loyalty.[12]

The exercise of power in an organization starts with the occupation of an organizational position. This provides a legitimate base on which to develop persuasive abilities, knowledge, interpersonal influence, and alliances which make legitimate power operable. At times it may be necessary to resort to the power granted to an organizational position from a higher echelon. But under these circumstances the influence and knowledge of the person play a role in making the formal power structure endure. Of what value is the concept of power to a manager? What are some ways in which a team manager can use power to influence the people with whom he works? John P. Kotter, writing in the *Harvard Business Review*, developed a series of characteristics that he found to be common in managers who were able to use power effectively in organizations.[13] According to him,

- Managers are sensitive to what other people consider to be legitimate behavior in getting and using power. To have power is to have obligations regarding its use and to behave as people expect them to behave.

---

[12] Peter F. Drucker, *Management: Tasks, Responsibilities, Practices* (New York: Harper & Row, 1974), p. 525.

[13] John P. Kotter, "Power, Dependence, and Effective Management," *Harvard Business Review*, July–August 1977, pp. 125–136.

- They understand the different types of power and influences and know what can be accomplished by each type at what costs and risk.
- Effective managers use all types of power and recognize that the right methods used in the right circumstances can improve organizational effectiveness.
- They look for managerial positions that give an opportunity to use power. Their career goals are consistent with the type of power they can use comfortably.
- They reinvest their resources in acquiring more power.
- Their use of power is mature and self-controlled. They avoid using power in a rash manner or for their own glorification.
- They are comfortable in using power to influence the lives and behavior of other people.

Kotter believes that as organizations grow more complex, managers are not able to accomplish their ends by using only persuasion and formal authority. They need to use power to influence people on whom they depend. He believes that managers create, increase, and maintain four different kinds of power:[14]

- By creating a sense of obligation in others, such as doing a favor for someone and expecting a favor in return.
- By building a reputation as an expert in certain matters, such as by publishing articles in professional journals, making speeches, etc.
- By fostering in others an unconscious identification with them or their ideas. A charismatic leader is able to do this. Consider the power that Martin Luther King developed.
- By reinforcing others' beliefs that they are dependent on them for assistance or protection. Control of money, equipment, and access to important people are examples of this type of power.

Power is something that everyone can develop an awareness of. According to a recently popular book: "We all have a power potential, but few of us use it, or even know it's there."[15]

To exercise effective total authority and power in an organizational position an individual must

- Have a well-defined delegation of authority from a legitimate source such as a higher-level organizational position

[14] Ibid.
[15] Michael Korda, *Power! How to Get it, How to Use It* (New York: Random House, 1975).

- Understand the nature of the job and how it relates to the other jobs and people with whom a person must deal.
- Develop the trust, confidence, and loyalty of the people with whom the individual works through the continued demonstration of the knowledge, skills, and attitudes required to be successful as a manager or professional in the organization.
- Develop an empathy with the professionals and peers in the creation of an environment where all concerned desire to work with economic, social, and psychological satisfaction in their reciprocal roles in the organization.

The power structure in modern organizations rests with those people who occupy a position and are vested with the authority attached to that position. Power is also exercised by those individuals who have specialized knowledge, skills, and an ability to rally people around them. This means that a manager cannot just tell people what to do—they expect to have a chance to "negotiate" fairly and openly about the conditions of the quality of life in the organization as well as about the nature of their work. Managers need to learn, or relearn as the case might be, how to be better negotiators. The art of negotiation is becoming a key skill of modern managers. Corporations, governments, unions, all types of organizations are applying negotiation in dealing with issues facing them. A book about negotiation, *Getting to Yes,* by Roger Fisher and William Ury at the Harvard Negotiation Project, has become a best-seller.[16]

Specialized consulting firms are offering seminars on the art of negotiation. Managers at all levels are learning the value of negotiations as a way of building supportive relationships in their organization and in so doing developing a power structure based on participative management, consensus decision making, and leadership by example.

Responsibility

Responsibility, a corollary of authority, is a state, quality, or fact of being responsible. A responsible person is one who is legally or ethically accountable for the care or welfare of other persons or an organization. A person who is responsible is expected to act without guidance or being told to do so by a superior authority. To be responsible is to be able to make rational decisions on one's own, to be trusted to make such decisions, and to be held accountable for one's decisions.

[16] Reported in Jeremy Main, "How to be a Better Negotiator," *Fortune*, September 19, 1983, pp. 141–146.

Accountability

Accountability is the state of assuming liability for something of value whether through a contract or because of one's position of responsibility. A professional is held responsible for excellence in the quality of the service rendered to the organization. A manager has dual accountability: Managers are held answerable for their own performance and for the performance of the people who report to them. One of the characteristics of a manager is to be held accountable for the effectiveness of the people who report to the manager.

Authority, responsibility, and accountability can rest with a single person or with a group of people. An example of pluralism in this sense is found in the use of the plural executive at the top-management level of organizations. The plural executive serves as an integrator of top-management decision making and implementation. Increasing complexity and size of many large organizations have created managerial responsibilities beyond the capabilities of one individual; the plural executive that has been created by organizations usually acts in an advisory capacity to the chief executive by providing stewardship for the strategic management of the company.

ORGANIZATIONAL ROLES[17]

Each individual plays a part in the organization to which he belongs. In the informal organization that role is assumed, subject to the continuing approval of the members of the informal organization. In the formal organization an individual is assigned a role reflecting the organizational position that the individual occupies and the knowledge and skills the individual brings to the organization.

When an individual assumes a position on a team, that position requires that a role be played to satisfy the authority, responsibility, and accountability of the position. Specific organizational roles require identification and negotiation, particularly in terms of the interfaces of individuals with superiors, peers, subordinates, and associates. Figure 5-2 illustrates the interdependencies that exist among the individuals. To be effective, everyone must understand his role and how to work with the others. These role interrelationships come to focus through work packages which are

[17] Portions of this section have been paraphrased from *Systems Analysis and Project Management*, 3rd Edition, by D. I. Cleland and W. R. King (McGraw-Hill Book Co., New York, 1983) and from "RIM Process in Participative Management," by D. F. Kocaoglu and D. I. Cleland, *Management Review*, October 1983, with permission.

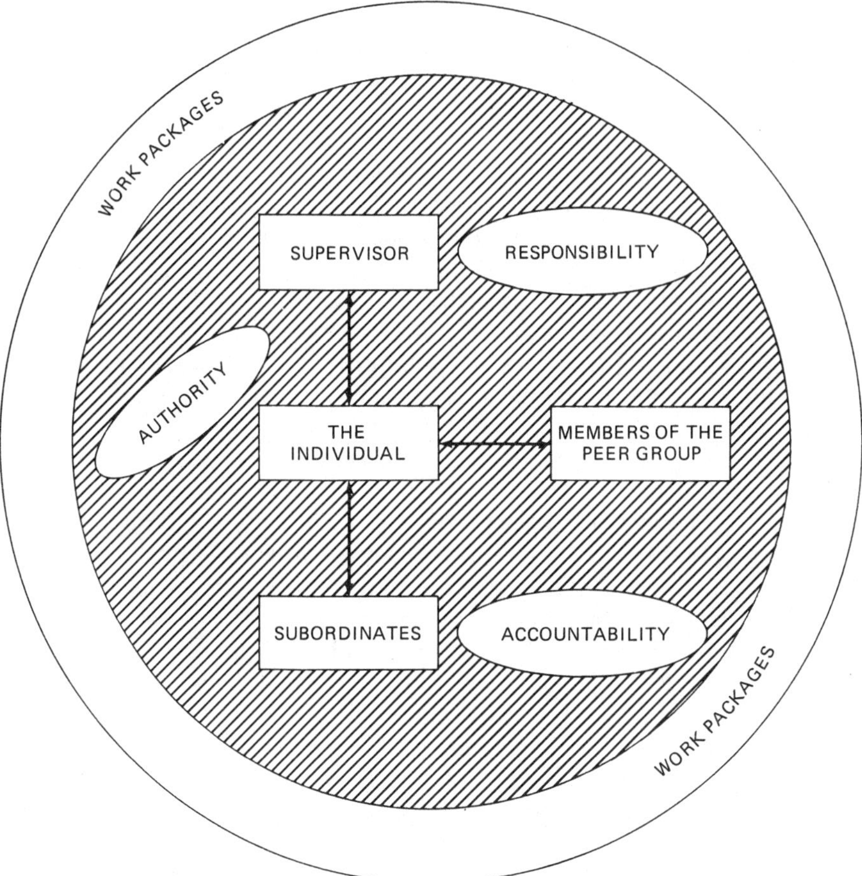

Figure 5-2. Role interdependencies.

major work elements at the hierarchical levels of the organization. Work packages are used to identify, assign, and control work in the organization; they have the following characteristics:

- A work package represents a discrete unit of work at the appropriate level of the organization where work is assigned.
- Each work package is clearly distinguished from all other work packages.
- The primary responsibility for completing the work package on schedule and within budget can always be assigned to an organizational unit or person and never to more than one unit or person.

- A work package can be integrated with other work packages at the same level of the organizational hierarchy to support the work packages at a higher level of the hierarchy.

In a manufacturing company, four major work packages at the top level of the organization would be found: production, finance, marketing, and research and development or engineering. Each of these work packages would be broken down into smaller and smaller units until the level of the lowest definable individual task is reached. Work packages exist vertically (within a functional specialty) and horizontally, that is, between functional specialties. The concept of the work package is consistent with the concept of authority, namely, identification of a person's role through delegation of authority for each work package down to the lowest level in the organization.

Work packages are level-dependent, becoming increasingly more general at each higher level, and increasingly more specific at each lower level. A president would be expected to have primary responsibility to set corporate objectives, whereas a contract manager would be expected to take the lead in developing corporate maintenance agreements for a profit center manager to use in supporting equipment that has been delivered to a customer. An individual is designated as having primary responsibility for each work package. Others who have collateral responsibility involving the work package are also designated. When these collective roles have been specified, there remains "no place to hide in the organization." If the work package is not completed on time or does not meet the performance standards, someone can readily be identified and held responsible and accountable for that work.

Responsibility, authority, and accountability—the triad of personal performance in organized life—are not left in doubt when individual and collective roles have been adequately defined. This definition can be carried out through the process of linear responsibility charting (LRC).

Linear Responsibility Charting*

An alternative to the traditional organization chart is found in the linear responsibility chart (LRC), which goes beyond a simple structural display of the formal alignment of the organization. An LRC portrays the specific relationship of an organizational position and a task, or work package, to be accomplished. Figure 5-3 shows an LRC. When several work packages

* Material in this section paraphrased in part from Dundar F. Kocaoglu and David I. Cleland, "RIM Process in Participative Management," *Management Review*, October 1983, and Chapter Eleven, D. I. Cleland and W. R. King, *Systems Analysis and Project Management*, 3rd Edition, McGraw-Hill Book Co., New York, 1983.

WORK PACKAGES	CONTRACT ADMIN.	MFG. MANAGER	FINANCIAL MANAGER	DIVISION MANAGER	R&D MANAGER	MARKETING MANAGER	MGMT. COUNCIL	COMMENTS
Setting Corporate Objectives	I,O	I,O	I,O	I,O	I,O	I,O	P	
Developing Corp. Maint. Agreements	P	N	I,N	I,A		I,N		
Negotiating Customer Contracts	I,R	I,N	R	P			A	
New Product Development		I,O	I	I*,O	P	I*,R		
Developing Bid Strategies	I,O	I,O		I,O		P		
Preparing the Annual Budget	I,O	I,O	P	I,O	I,O	I,O	A	
Developing Master Sched. for Opers.	N	P	N	I,N	I,N			
Establishing Standard Costs		W	P	I,O				

Legend:
- (P) Primary responsibility—the prime authority and responsibility for accomplishment of work package
- (R) Review—reviews output of work package
- (N) Notification—is notified of output of work package
- (A) Approval—approves work package
- (O) Output—receives output
- (I*) Initiation—work package initiated by incumbent
- (I) Input—provides input to work package
- (W) Work is done—accomplishes actual labor of work package

P includes W
P includes A
P include I*
A includes R
Unless otherwise specified

Figure 5-3. Sample linear responsibility chart.

are involved, one of the key advantages of using an LRC to establish organizational relationships becomes apparent, that is, the people are brought into a dialog as to their specific authority and responsibility to a work package and to other individuals in the organization.

The LRC has three key elements:

1. Organizational positions taken from the traditional organizational chart and used as the column headings in the LRC.
2. Work packages defined at all levels and used as the rows in the LRC.
3. Work-position relationships (WPRs) specified at the interface points and located at the intersections of the organizational positions and the work packages. These are the points where activities are carried out to make the organization function effectively.

The major WPRs that should be given utmost attention are discussed in the following paragraphs.

P—Primary Responsibility: Prime authority, responsibility, and accountability for the accomplishment of the work package rests in this position. It implies leadership for the work package to see that the work is accomplished on time, within budget, and meeting the performance standards. The person who has primary responsibility is "it" for doing what needs to be done. Each work package has one and only one P associated with it. No work can be conducted without assigning the primary responsibility to an organizational position; and the primary responsibility for a work package cannot be assigned to more than one position at a time. This follows the definition of a work package, as a discrete element of work at a given organizational level, for which the primary responsibility is uniquely identified.

A—Approval: This organizational position approves the work package. A is used when the authentication of a work package is required, as in the case of the approval of a capital investment project by the chief executive officer of a corporation, for example.

R—Review: The individual who has an R reviews the output of the work package to decide if the predetermined quality and/or quantity standards have been met. For example, a financial executive reviews the financial credibility of the bid price on a proposal submitted to the customer, and a quality control manager reviews the outputs of the production line. Typically, A includes R unless otherwise specified. The reason for this is the fact that if the review of a work package is not explicitly assigned to an organizational unit, then the approval authority carries with it the review function responsibility.

N—Notification: The individual occupying the position with an N is notified of the output of the work package. As a result of this notification, he then decides whether or not any action should be taken. For example, when the sales department bids on a request for proposal, the contract administrator is notified of the decision and he, in turn, takes the contracts-related legal precautions as needed.

O—Output: This position receives the output of the work package and integrates it into the work being accomplished. For example, the contract administrator receives a copy of the engineering change orders so that the effects of changes on the terms and conditions of a project contract can be determined. There is a subtle difference between N and O. N (notification) can be a telephone call, a memo, or a short verbal notice, whereas O (output) implies that the final product or service resulting from the work package is delivered to the organizational position thus identified.

I—Input: This position provides input to the work package. Input is generally required from various positions in the organization. For example, a bid/no bid decision on a contract cannot be made by a company unless inputs are received from the manufacturing manager, financial manager, contract administrator, marketing manager, and the manager of the profit center division, which will eventually be responsible for that contract.

W—Work is done: The actual labor of the work package is accomplished by the position with a W. This legend is used when the work is done by a position other than where the primary responsibility resides. For example, the primary responsibility for the preparation of a project proposal normally resides with the sales manager, but the preparation of the document itself is accomplished by the publications manager in the organization.

I—Initiation:* The work package is initiated by the position marked with I*. For example, new product development is the responsibility of the R&D manager, but the process is generally initiated with a request either from the profit center manager or from the marketing manager. For this work package, the R&D manager has a P, the profit center manager or the marketing manager has an I*.

If W, A, R, and I* are not separately identified, then P is assumed to include them.

In Figure 5-3, a sample LRC in an actual industrial organization, management council is included as the "plural executive," composed of the key managers of the organization, dealing with key operational and strategic issues.

Each column of the LRC describes how a certain organizational position operates in the company. It summarizes the line, staff, and support roles and the information dissemination characteristics of the position. For example, in the sample LRC, the R&D manager has primary responsibility for the development of new products, but plays only a support role for the other work packages. He provides input to the financial manager for annual budget and another input to the manufacturing manager for master schedules. In turn, he receives the corporate objectives from the management council, and the final budget from the financial manager. He is also notified by the manufacturing manager when the master schedule is completed.

In a similar way, the rows contain information about how certain jobs are accomplished, and who does what for whom in carrying out the work package requirements. For example, bid strategies in the sample LRC are developed by the marketing manager with inputs from the contract administrator, the manufacturing manager, and the profit center division manager. After the strategies are developed, each of those managers receives a document from the marketing manager, describing the policies on how the bids are to be prepared and submitted.

Taken as a whole, LRC is a blueprint of the activity and information flows that take place in the organizational interfaces in a company. The authority/responsibility patterns and the organizational interdependencies can be read directly from the chart. Once the LRC is developed, it can be sorted for each organizational position, first with all the P's, then with I's, O's, and so on. When the manager looks at the sorted work packages related to his organizational unit, he immediately sees a listing of the activities for which he has direct responsibility and those for which he supports the other units in the organization. A study of the WPRs (work/position relationships) vertically gives him his position description. Basically, it is a delineation of what he does for each work package. If he looks at the WPRs horizontally, he sees whom he interacts with and how. The LRC identifies his contact points in the organization, and the nature of contracts he is to maintain.

The LRC is a valuable tool as succinct description of organizational interfaces. It conveys more information than several pages of job descriptions and policy documents by delineating the authority/responsibility relationships and specifying the accountability of each organizational position. However, by far the most important aspect of the LRC is the process through which the people in the organization prepare it. If the LRC is developed in an autocratic fashion, it simply becomes a document portraying the organizational relationships. On the other hand, if it is prepared through a participative process, the final output becomes secondary to the impacts of the process itself. The open communications, broad discussions, resolution of conflicts, and achievement of consensus through participation provide a solid basis for organizational development and managerial harmony. By the time the LRC is developed this way, the organization goes through such an "education" that the chart becomes secondary. Organizational units fully subscribe to the ideas behind the LRC and protect its integrity because of their vested interests amplified by the participative process.

Developing the LRC

The development of the LRC is inherently a group activity, involving the key people in the organization who have some vested interest in the work that is to be done.

The following plan for the development of an LRC should prove useful:

1. Distribute copies of the current traditional organization chart and position descriptions to the key people.
2. Develop and distribute blank copies of the LRC.
3. At the first opportunity, get the people together to discuss
   a. The advantages and shortcomings of the traditional organization chart.
   b. The concept of a project work breakdown structure and the resulting work packages.
   c. The nature of the linear responsibility chart, how it is developed, and how it is used.
   d. A way of establishing a symbology to show the work-package-organization-position relationship. Getting a meeting of the minds on this symbology is very important so that the people will be willing to commit themselves to such a relationship.
   e. The makeup of the actual work breakdown structure with accompanying work packages.
   f. The fitting of the symbols into the proper relationship in the LRC.
4. Encourage an intensive dialog during the actual making of an LRC. In such a meeting people will tend to be protective of their organizational territory. The LRC by its nature requires a commitment to support and share the allocation of organizational resources applied to work packages. This commitment requires the ability to communicate and decide. The process simply takes time. But it is worth the effort: When the LRC is completed, the people are much more knowledgeable about what is expected of them.

Building the Team

The use of the LRC process in determining relative roles in the organization creates a very high degree of the awareness of teamwork that goes on in the day-to-day activities of the organization. The LRC shows that interdependence is a fact of life in organizational teams. This interdependence operates both vertically and horizontally; superiors and subordinates are dependent on each other and on their peers in the conduct of the organization's affairs. Subordinates relay information and comply with the manager's orders. The boss has information that the subordinate needs concerning his own job. When an individual has to coordinate and work with peers to obtain information or compliance with a procedural requirement, an interdependency relationship exists. For a well-functioning organization, these relationships come together in effective teamwork.

Such teamwork is subtle. As managers and professionals work together on reciprocal work packages, many of the cultural characteristics found in a well-functioning team become visible:

- Managers work at integrating work packages through information sharing and decision making.
- Subordinates share information with others and build a communication network centered around the work packages that need to be accomplished.
- An ambience of trust, loyalty, confidence, and commitment develops as managers share resources, views, results, and decisions.
- Managers become better problem solvers thanks to more communication and peer support. A collective strength is formed that is far greater than would be realized with one manager working alone.
- More creativity and innovation emerge as managers and professionals share ideas (and problems) with others. Having a small peer group off whom one can bounce ideas helps to develop ideas into worthwhile concepts.

When the team culture becomes established, conflict becomes socially acceptable. The LRC process provides one way of getting conflict over organizational territory out into the open for discussion and eventual resolution. As noted by one author:

> We increasingly understand that psychological and social energy is tied up in suppressing conflict, that conflicts not confronted may be played out in indirect and destructive ways, and that the differences that underlie interpersonal conflict often represent diversity or complementarity of significant potential value to the organization. An interpersonal or organizational system that can acknowledge and effectively confront its internal conflicts has a greater capacity to innovate and adapt.[18]

[18] R. Walton, *1969 Interpersonal Peacemaking: Confrontations and Third Party Consultation* (Reading, Mass.: Addison-Wesley, 1969).

Conflict is a natural form of behavior among the team members when they have different primary responsibilities, interests, perceptions, and attitudes on how to handle issues. When the people have had an opportunity to participate in developing the specificity of their organizational roles through charting of the LRC, and know what is expected of them, conflict can be accepted and worked through for a solution.

The LRC is a process that helps people to understand how they relate to the organizational work packages. The process of developing the LRC is as important as the product—the completed LRC. Once developed, it becomes a model for the formal intended relationships in the enterprise. Once assembled, the LRC can become a "living document" to be used for the following:

- Portray formal authority, responsibility, accountability relationships
- Acquaint newcomers with how things are done on the team
- Get people committed and motivated—they know specifically what is expected of them
- Bring out real or potential conflict over territorial prerogatives in the organization
- Permit people to see the "big picture"—how they fit into the larger whole
- Facilitate teamwork—people have greater opportunity to see their role on the team
- Provide standard against which the supervisor can monitor what people are doing
- Set the stage for further participation in the organization—assuming that the LRCs are developed in a participative fashion
- Encourage communication about work—people have a basis on which to discuss their work with superiors, subordinates, peers, and associates
- Help to take a change in organization structure and translate it into specific roles for individuals

Alternative Applications of the LRC Process

The LRC can be applied in different situations. Examples of these applications follow.

Blending the LRC and the work package together can help in other ways besides establishing organizational relationships. A task force in one company engaged in relocating from California to the East Coast identified all of the work packages that would be involved and put these work packages on an LRC to assign responsibility and on a Gantt chart to establish a timetable when the move of the work package would be completed. An excerpt from one of the Gantt charts is shown in Figure 5-4.

Figure 5-5 shows how the standard LRC format can be converted into a report distribution matrix for both in-house personnel and customers. Figure 5-6 is most often found in project plans and proposals for construction activities. In another application, an inventory skills matrix, the names of all project office (or project team) personnel are shown across the top, and listed on the side are the specific functions or tasks to be performed. The inventory skill matrix accomplishes three functions. First, it shows the customer the background skills of the employees. Second, it shows the customer that each major function or task is accounted for. Third, and most important, it shows the customer the *backup* skills on the project such that if one or more people leave the project, there is still sufficient skill on the team to compensate for the loss of personnel.

WORK PACKAGES	JAN 25	FEB 1 8 15 22	MAR 1 8 15 22 29	APRIL 5 12 19 26	MAY 3 10 17 24 31	JUNE 7 14 21 28	JULY 5 12 19 26
Hardware and Software Training; Drafting Equipment	△22	▲8					
East Coast Customer Files and Sales Literature		22△	8▲				
Accounting and Administration Equipment				1△ ▲12			
5000/6200 RTU Dwgs; Master Station Subassem and Con/Rack and Console		22△	▲15				
7200 RTU FAB, PCB Dwg; Rack and Subassemblies			29△	▲12			
Firmware Documentation		22△	15▲				
Manufacturing Equipment (partial)		22△	15▲				
Manufacturing Equipment				26△	▲10		

Figure 5-4. LRC Gantt chart.

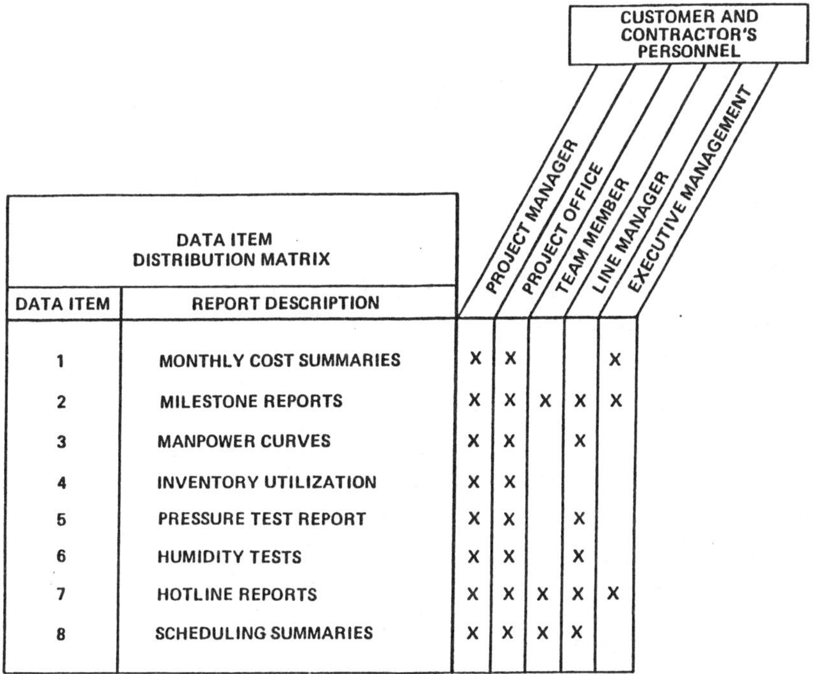

DATA ITEM	REPORT DESCRIPTION	PROJECT MANAGER	PROJECT OFFICE	TEAM MEMBER	LINE MANAGER	EXECUTIVE MANAGEMENT
1	MONTHLY COST SUMMARIES	X	X			X
2	MILESTONE REPORTS	X	X	X	X	X
3	MANPOWER CURVES	X	X		X	
4	INVENTORY UTILIZATION	X	X			
5	PRESSURE TEST REPORT	X	X		X	
6	HUMIDITY TESTS	X	X		X	
7	HOTLINE REPORTS	X	X	X	X	X
8	SCHEDULING SUMMARIES	X	X	X	X	

Figure 5-5. Report distribution matrix.

FUNCTIONAL AREAS OF EXPERTISE / PROJECT TEAM	ABLE, J.	BAKER, P.	COOK, D.	DIRK, L.	EASLEY, P.	FRANKLIN, W.	GREEN, C.	HENRY, L.	IMHOFF, R.	JULES, C.	KLEIN, W.	LEDGER, D.	MAYER, Q.	NEWTON, A.	OLIVER, G.	PRATT, L.
ADMINISTRATIVE MANAGEMENT		a				a		a			a	a			a	
COST CONTROL		b	b		b	b	b				b	b		b	b	
ECONOMIC ANALYSIS	c			c				c	c				c			c
ENERGY SYSTEMS		d	d		d		d			d			d		d	d
ENVIRONMENTAL IMPACT ASSESSMENT	e	e	e						e		e		e			
INDUSTRIAL ENGINEERING	f				f					f						
INSTRUMENTATION	g			g		g					g				g	
PIPING AND DESIGN LAYOUT	h		h		h	h				h			h			
PLANNING AND SCHEDULING		i		i	i			i				i		i		i
PROJECT MANAGEMENT	j			j		j					j				j	
PROJECT REPORTING		k	k		k			k	k			k		k		k
QUALITY CONTROL		l	l			l	l	l	l							
SITE EVALUATION		m				m			m	m				m		
SPECIFICATION PREPARATION			n	n			n				n		n			n
SYSTEMS DESIGN		o	o		o		o	o		o		o			o	

Figure 5-6. Personnel skills matrix.

SUMMARY

A team leader is expected to take the initiative in developing a sense of direction for the team's work in its future. The result of planning is a schema for how the team will go about meeting its future. Getting the members of the team organized is an important team leader's responsibility.

The linear responsibility chart is a tool for participative management in developing an understanding of roles in organizations. If used properly, it provides valuable assistance in getting the managerial team organized and in pushing the management responsibility to the lowest levels of the organization. It helps to develop open communications, mutual trust, and confidence among the organizational units working toward common goals and objectives.

The resulting matrix is a simple pictorial representation of the work flows and responsibility interfaces in the organization. But the process through which the matrix is developed is much more critical. It is through that continuous process that the organizational interactions become visible; the territorial fears are subdued; the roles, responsibilities, authorities, and accountabilities are crystallized; and the entire organization is pulled together toward the achievement of its mission.

The LRC facilitates the practice of teamwork in organizations. The resulting teamwork is tied together in a sharing context of decision making, results, and accountability. Within such a context the professional has greater ability to be heard and influence the course of events in the organization.

Conflict in team management is discussed in Chapter 6.

BIBLIOGRAPHY

Abramson, B. N., and R. D. Kennedy. *Managing Small Projects*. TRW Systems Group, TRW Inc., 1969.

Cleland, D. I., and W. R. King. *Systems Analysis and Project Management*, 3rd ed. New York: McGraw-Hill, 1983.

Drucker, P. F. *Management: Tasks, Responsibilities, Practices*. New York: Harper & Row, 1974.

Fayol, Henri. *General and Industrial Management*. London: Sir Isaac Pitman & Sons, Ltd., 1949.

Kocaoglu, D. F., and D. I. Cleland. "RIM Process in Participative Management." *Management Review*, October 1983.

Korda, Michael. *Power! How to Get It, How to Use It*. New York: Random House, 1975.

Kotter, John P. "Power, Dependence, and Effective Management." *Harvard Business Review*, July–August 1977.

Main, Jeremy. "How to Be a Better Negotiator." *Fortune*, September 19, 1983.

Thompson, James D. "Authority and Power in Identical Organizations." *The American Journal of Sociology,* vol. 60, November 1956.

Walton, R. *1969 Interpersonal Peacemaking: Confrontations and Third Party Consultation.* Reading, Mass.: Addison-Wesley, 1969.

"X Cars Exit." *Fortune,* February 21, 1983.

6
CONFLICTS[1]

INTRODUCTION

Opponents of such team management structures as project management assert that the major reason why many companies avoid changeover to a project management organizational structure is either fear or inability to handle the resulting conflicts. Conflicts are a way of life in a team structure and can generally occur at any level in the organization, usually the result of conflicting objectives. In this chapter we will discuss the subject of conflict in the management of teams. The project team will be used as a demonstration model for the discussion of conflict in team management.

The team manager has often been described as a conflict manager. In many organizations the team manager continually fights fires and crises evolving from conflicts, and delegates the day-to-day responsibility of running the project to the project team members. Although this is not the best situation, it cannot always be prevented from occurring, especially after organizational restructuring or the initiation of projects requiring new resources.

The ability to handle conflicts requires an understanding of why conflicts occur. Four questions can be asked, the answers to which should be beneficial in handling, and possibly preventing, conflicts.

1. What are the project objectives and can they be in conflict with other projects?
2. Why do conflicts occur?
3. How do we resolve conflicts?
4. Is there any type of preliminary analysis that could identify possible conflicts before they occur?

---

[1] The majority of this chapter has been taken from H. Kerzner, *Project Management: A Systems Approach to Planning, Scheduling and Controlling* (New York: 2nd ed. Van Nostrand Reinhold, 1984), Chapter 7.

THE CONFLICT ENVIRONMENT

In the project environment, conflicts are inevitable. However, conflicts and their resolution can be planned for. For example, conflicts can easily develop out of a situation where members of a team have a misunderstanding of each other's roles and responsibilities. Through documentation, such as the linear responsibility charts, it is possible to establish formal organizational procedures (either at the project level or companywide). Resolution means collaboration in which people must rely upon one another. Without this, mistrust will prevail and activity documentation can be expected to increase.

The most common type of conflicts involve

- Manpower resources
- Equipment and facilities
- Capital expenditures
- Costs
- Technical opinions and trade-offs
- Priorities
- Administrative procedures
- Scheduling
- Responsibilities
- Personality clashes

Each of these conflicts can vary in relative intensity over the life cycle of a project. The relative intensity can vary as a function of

- Getting closer to project constraints
- Having only two constraints instead of three (i.e., time and performance, but not cost)
- The project life cycle itself
- The person whom the conflict is with

Sometimes conflict is "meaningful" and produces beneficial results. These meaningful conflicts should be permitted to continue as long as project constraints are not violated and beneficial results are being received. An example of this would be two technical specialists arguing that each has a better way of solving a problem, and each trying to find additional supporting data for his hypothesis.

Some conflicts are inevitable and continuously reoccur. As an example, let us consider the raw material and finished goods inventory. Manufacturing wants the largest possible inventory of raw materials on hand so as not to shut down production; sales and marketing want the largest finished goods inventory so that customer demands will be met; and, finally, finance and accounting want the smallest raw material and finished goods inventory so the books will look better and no cash flow problems will occur.

Conflicts appear differently depending on the organizational structure. In the traditional structure, conflict should be avoided; in the project structure, conflict is part of change and therefore inevitable. In the traditional structure, conflict is the result of troublemakers and egoists; in the project structure, conflict is determined by the structure of the system and relationship among components. In the traditional structure, conflict is bad; in the project structure, conflict may be beneficial.

Conflicts can occur with anyone and over anything. Some people contend that personality conflicts are the most difficult to resolve. Below are several situations. The reader might consider what he or she would do if placed in the situations.

- Two of your functional team members appear to have personality clashes and almost always assume opposite points of view during decision making. They are both from the same line organization. Conflicts are inevitable.
- Two of your line managers continuously argue as to who should perform a certain test. You know that this situation exists, and that the department managers are trying to work it out themselves, often with great pain. However, you are not sure for how long they will be able to resolve the problem themselves.
- Manufacturing says that they cannot produce the end item according to engineering specifications.
- R&D quality control and manufacturing operations quality control argue as to who should perform a certain test on an R&D project. R&D postulates that it is their project, and manufacturing argues that it will eventually go into production and that they wish to be involved as early as possible.
- During contract negotiations, a disagreement occurs. The vice-president of Company A orders his director of finance, the contract negotiator, to break off negotiations with Company B because the contract negotiator for Company B does not report directly to a vice-president.
- Mr. X is the project manager of a $65 million project of which $1 million is subcontracted out to another company in which Mr. Y is the project manager. Mr. X does not consider Mr. Y his counterpart and continuously communicates with the director of engineering in Mr. Y's company.

Ideally, the project manager should report high enough so that he can get timely assistance in resolving conflicts. Unfortunately, this is easier said than done. Therefore, project managers must plan for conflict resolution, as shown in the following examples:

- The project manager might wish to concede on a low-intensity conflict if he knows that a high-intensity conflict is expected to occur at a later point in the project.
- Jones Construction Company has recently won a $120 million effort for a local company. The effort includes three separate construction projects, each one beginning at the same time. Two of the projects are 24 months in duration, and the third one is 36 months. Each project has its own project manager. When resource conflicts occur between the projects, the customer is usually called in.
- Susan is a department manager who must supply resources to four different projects. Although each project has an established priority, the project managers continuously argue that departmental resources are not being allocated effectively. Susan now holds a monthly meeting with all four of the project managers and lets them determine how the resources should be allocated.

Many executives feel that the best way of resolving conflicts is by establishing priorities. This may be true as long as priorities are not continuously shifted around. As an example, Minnesota Power and Light establishes priorities as follows:

- Level 0: no completion date
- Level 1: to be completed on or before a specific date
- Level 2: to be completed on or before a given fiscal quarter
- Level 3: to be completed within a given year

This type of technique will work as long as we do not have a large number of projects in any one group, say level 1. How would we then distinguish between projects?

Executives are responsible for establishing priorities and often make the mistake of *not* telling the project managers the reasons for the priority level. There may be sound reasons for concealing this information, but this practice should be avoided whenever possible.

The most common factors influencing the establishment of project priorities include the following:

- The technical risks in development
- The risks that the company will incur, financially or competitively

- The nearness of the delivery date and the urgency
- The penalties that can accompany late delivery dates
- The expected savings, profit increase, and return on investment
- The amount of influence that the customer possesses, possibly due to the size of the project
- The impact on other projects
- The impact on affiliated organizations
- The impact on a particular product line

The ultimate responsibility for establishing priorities rests with top-level management. Yet even with priority establishment, conflicts still develop. David Wilemon has identified several reasons why conflicts still occur:[2]

- The greater the diversity of disciplinary expertise among the participants of a project team, the greater the potential for conflict to develop among members of the team.
- The lower the project manager's degree of authority, reward, and punishment power over those individuals and organizational units supporting his project, the greater the potential for conflict to develop.
- The less the specific objectives of a project (cost, schedule, and technical performance) are understood by the project team members, the more likely it is that conflict will develop.
- The greater the role ambiguity among the participants of a project team, the more likely it is that conflict will develop.
- The greater the agreement on superordinate goals by project team participants, the lower the potential for detrimental conflict.
- The more the members of functional areas perceive that the implementation of a project management system will adversely ursurp their traditional roles, the greater the potential for conflict.
- The lower the percent need for interdependence among organizational units supporting a project, the greater the potential for dysfunctional conflict.
- The higher the managerial level within a project or functional area, the more likely it is that conflicts will be based upon deep-seated parochial resentments. By contrast, at the project or task level, it is more likely that cooperation will be facilitated by task orientation and professionalism that a project requires for completion.

---

[2] David L. Wilemon, "Managing Conflict in Temporary Management Situations." *The Journal of Management Studies*, 1973, pp. 282–296.

MANAGING CONFLICT

Temporary management situations produce conflicts. This is a natural occurrence resulting from the differences in the organizational behavior of individuals, the differences in the way that functional and project managers view the work required, and the lack of time necessary for project managers and functional personnel to establish ideal working relationships.

Regardless of how well planning is developed, project managers must be willing to operate in an environment that is characterized by constant and rapid change. This turbulent environment can be the result of changes in the scope of work, a shifting of key project and functional personnel due to new priorities, and other unforeseen developments. The success or failure of a project manager is quite often measured by the ability to deal with change.

> In contrast to the functional manager who works in a more standardized and predictable environment, the project manager must live with constant change. In his effort to integrate various disciplines across functional lines, he must learn to cope with the pressures of the changing work environment. He has to foster a climate that promotes the ability of his personnel to adapt to this continuously changing work environment. Demanding compliance to rigid rules, principles, and techniques is often counter-productive. In such situations, an environment conducive to effective project management is missing and the project leader too often suffers the same fate as heart-transplant patients—rejection![3]

There is no one single method that will suffice for managing all conflicts in temporary management situations because

- There exist several types of conflicts.
- Each conflict can assume a different relative intensity over the life cycle of the project.

The detrimental aspects of these conflicts can be minimized if the project manager can anticipate their occurrence and understand their composition. The prepared manager can then resort to one of several conflict resolution modes in order to more effectively manage the disagreements that can occur.[4]

---

[3] H. S. Dugan, H. J. Thamhain, and D. L. Wilemon, "Managing Change in Project Management," *Proceedings of the Ninth Annual International Seminar/Symposium on Project Management,* Chicago, October 22–26, 1977, pp. 178–188.

[4] The remainder of this section is devoted to Hans J. Thamhain and David L. Wilemon, "Conflict Management in Project Life Cycles," *Sloan Management Review,* Summer 1975, pp. 31–50. Reprinted by permission.

Thamhain and Wilemon surveyed 150 project managers on conflict management. Their research tried to determine the type and magnitude of the particular type of conflict which is most common at specific life-cycle stages, regardless of the particular nature of the project. For the purpose of their paper the authors stated the following definitions:

> Conflict is defined as the behavior of an individual, a group, or an organization which impedes or restricts (at least temporarily) another party from attaining its desired goals. Although conflict may impede the attainment of one's goals, the consequences may be beneficial if they produce new information which, in turn, enhances the decision-making process. By contrast, conflict becomes dysfunctional if it results in poor project decision-making, lengthy delays over issues which do not importantly affect the outcome of the project, or a disintegration of the team's efforts.[5]

The study presented in their paper was part of an ongoing and integrated research effort on conflict in the project-oriented work environment.[6–9]

Project managers frequently indicate that one of the requirements for effective performance is the ability to effectively manage various conflicts and disagreements that invariably arise in task accomplishment. While several research studies have reported on the general nature of conflict in project management, few studies have been devoted to the cause and management of conflict in specific project life-cycle stages. If project managers are aware of some of the major causes of disagreements in the various project life-cycle phases, there is a greater likelihood that the detrimental aspects of these potential conflict situations can be avoided or minimized.

This study first investigates the mean intensity of seven potential conflict determinants frequently thought to be prime causes of conflict in project management. Next, the intensity of each conflict determinant is

---

[5] H. J. Thamhain and D. L. Wilemon, "Conflict Management in Project-Oriented Work Environments," *Proceedings of the Sixth International Meeting of the Project Management Institute,* Washington, D.C., September 18–21, 1974.

[6] "Diagnosing Conflict Determinants in Project Management," *IEEE Transactions on Engineering Management,* **22** (1975):35–44.

[7] D. L. Wilemon and J. P. Cicero, "The Project Manager—Anomalies and Ambiguities," *Academy of Management Journal,* Fall 1970, pp. 269–282.

[8] D. L. Wilemon, "Project Management Conflict: A View From Apollo," *Proceedings of the Third Annual Symposium of the Project Management Institute,* Houston, Texas, October 1971.

[9] D. L. Wilemon, "Project Management and its Conflicts: A View From Apollo," *Chemical Technology,* **2**(9) (1972):527–534.

viewed from the perspective of individual project life-cycle stages. An examination is then made of various conflict-handling modes used by project managers which leads to a number of suggestions for minimizing the detrimental effects of conflict over the project life cycle.

Research Design

Approximately 150 managers from a variety of technology-oriented companies were asked to participate in this comprehensive research project. A usable sample of 100 project managers was eventually selected for this study.

A questionnaire was used as the principal data collection instrument. In addition, discussions were held with a number of project managers on the subject under investigation to supplement the questionnaire data and the resulting conclusions. This process proved helpful in formulating a number of recommendations for minimizing detrimental conflicts.

The development of the questionnaire relied on several pilot studies. It was designed to measure values on three variables (1) the average intensity of seven potential conflict determinants over the entire project life cycle; (2) the intensity of each of the seven conflict sources in the four project life-cycle phases; and (3) the conflict resolution modes used by project managers.

Mean Conflict Intensity

The average conflict intensity perceived by the project managers was measured for various conflict sources and for various phases of the project life cycle. Project managers were asked to rank the intensity of conflict they experienced for each of seven potential sources on a standard four-point scale. The seven potential sources are as follows:

1. *Conflict over project priorities:* The views of project participants often differ over the sequence of activities and tasks that should be undertaken to achieve successful project completion. Conflict over priorities may occur not only between the project team and other support groups but also within the project team.
2. *Conflict over administrative procedures:* A number of managerial and administrative-oriented conflicts may develop over how the project will be managed; that is, the definition of the project manager's reporting relationships, definition of responsibilities, interface relationships, project scope, operational requirements, plan of exe-

cution, negotiated work agreements with other groups, and procedures for administrative support.

3. *Conflict over technical opinions and performance trade-offs:* In technology-oriented projects, disagreements may arise over technical issues, performance specifications, technical trade-offs, and the means to achieve performance.
4. *Conflict over manpower resources:* Conflicts may arise around the staffing of the project team with personnel from other functional and staff support areas or from the desire to use another department's personnel for project support even though the personnel remain under the authority of their functional or staff superiors.
5. *Conflict over cost:* Frequently, conflict may develop over cost estimates from support areas regarding various project work breakdown packages. For example, the funds allocated by a project manager to a functional support group might be perceived as insufficient for the support requested.
6. *Conflict over schedules:* Disagreements may develop around the timing, sequencing, and scheduling of project-related tasks.
7. *Personality conflict:* Disagreements may tend to center on interpersonal differences rather than on "technical" issues. Conflicts often are "ego-centered."

Intensity of Specific Conflict Sources by Project Life-Cycle Stage

The conflict intensity experienced by project managers for each source over the four life-cycle stages was measured on a special grid. The x-axis of the grid identifies four standard life-cycle phases: project formation, project build-up, main program phase, and phaseout. The y-axis delineates the seven potential sources of conflict. The respondents were asked to indicate on a standard four-point scale the intensity of the conflict they experienced for each of the seven potential sources of conflict within each of the four project life-cycle stages.

Conflict-Handling Modes

A number of research studies indicate that managers approach and resolve conflicts by utilizing various conflict resolution modes. Blake and Mouton,[10] for example, have delineated five modes for handling conflicts:

[10] R. R. Blake and J. S. Mouton, *The Managerial Grid* (Houston: Gulf Publishing, 1964).

1. *Withdrawal:* Retreating or withdrawing from an actual or potential disagreement.
2. *Smoothing:* De-emphasizing or avoiding areas of difference and emphasizing areas of agreement.
3. *Compromising:* Bargaining and searching for solutions that bring some degree of satisfaction to the parties in a dispute. Characterized by a "give-and-take" attitude.
4. *Forcing:* Exerting one's viewpoint at the potential expense of another. Often characterized by competitiveness and a win/lose situation.
5. *Confrontation:* Facing the conflict directly which involves a problem-solving approach whereby affected parties work through their disagreements.[11]

Aphorisms or statements of folk wisdom were used as surrogates for each conflict resolution mode.[12] The project managers were asked to rank the accuracy of each proverb in terms of how accurately it reflected the actual way in which they handled disagreements in the project environment. Fifteen proverbs were selected to match the five conflict-handling modes identified by Blake and Mouton.[13,14] This analysis provides an insight into the perceived conflict-handling mode of the project managers.

Analysis of Results

The results of the study are presented in three parts.

Mean Conflict Intensity over the Project Life Cycle

The mean intensity experienced for each of the potential conflict sources over the entire life of projects is presented in Figure 6-1. As indicated,

---

[11] For a full description of these definitions, see R. J. Burke, "Methods of Resolving Interpersonal Conflict," *Personnel Administration*, July–August 1969, pp. 48–55. Also see H. J. Thamhain and D. L. Wilemon, "Conflict Management in Project-Oriented Work Environments," *Proceedings of the Sixth International Meeting of the Project Management Institute*, Washington, D.C., September 18–21, 1974.

[12] Specifically, the measurements rely on the research of P. R. Lawrence and J. W. Lorsch, "New Management Job: The Integrator," *Harvard Business Review*, November–December 1967, pp. 142–152.

[13] See footnote 10.

[14] These proverbs have been used in other research of a similar nature to avoid the potential bias that might be introduced otherwise by the use of social science jargon. For further details, see R. J. Burke, "Methods of Managing Superior-Subordinate Conflict," *Canadian Journal of Behavioral Science*, **2**(2) (1970):124–135.

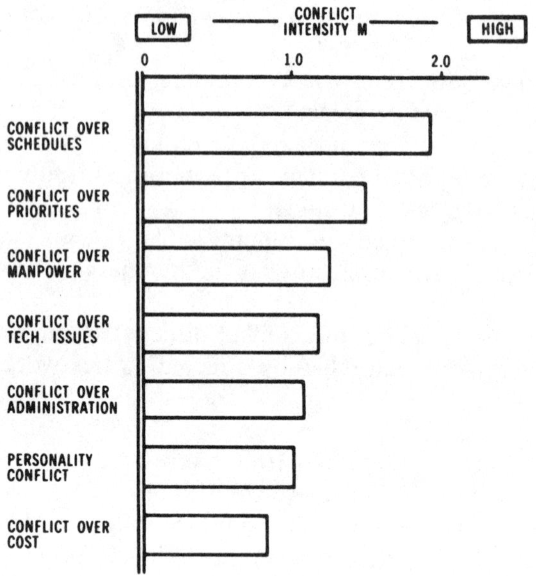

Figure 6-1. Mean conflict intensity profile over project life cycle. (*Source:* Hans J. Thamhain and David L. Wilemon, "Conflict Management in Project Life Cycles," *Sloan Management Review,* Summer 1975, pp. 31–50. Reprinted by permission.)

relative to other situations, disagreements over schedules result in the most intense conflict over the total project. Scheduling conflicts often occur with other support departments over whom the project manager may have limited authority and control. Scheduling problems and conflicts also often involve disagreements and differing perceptions of organizational departmental priorities. For example, an issue urgent to the project manager may receive a low-priority treatment from support groups and/or staff personnel because of a different priority structure in the support organization. Conflicts over schedules frequently result from the technical problems and manpower resources.

Conflict over project priorities ranked second highest over the project life cycle. In our discussions with project managers, many indicated that this type of conflict frequently develops because the organization did not have prior experience with a current project undertaking. Consequently, the pattern of project priorities may change from the original forecast, necessitating the reallocation of crucial resources and schedules, a process that is often susceptible to intense disagreements and conflicts. Similarly, priority issues often develop into conflict with other support departments whose established schedules and work patterns are disturbed by the changed requirements.

Conflict over manpower resources was the third most important source of conflict. Project managers frequently lament when there is little "organizational slack" in terms of manpower resources, a situation in which they often experience intense conflicts. Project managers note that most of the conflicts over personnel resources occur with those departments who either assign personnel to the project or support the project internally.

The fourth strongest source of conflict involved disagreements over technical opinions and trade-offs. Often the groups who support the project are primarily responsible for technical inputs and performance standards. The project manager, on the other hand, is accountable for the cost, schedule, and performance objectives. Since support areas are usually responsible for only parts of the project, they may not have the broad management overview of the total project. The project manager, for example, may be presented with a technical problem. Often he must reject the technical alternative owing to cost or schedule restraints. In other cases, he may find that he disagrees with the opinions of others on strictly technical grounds.

Conflict over administrative procedures ranked fifth in the profile of seven conflict sources. Most of the conflict over administrative procedures that occurs is almost uniformly distributed with functional departments, project personnel, and the project manager's superior.[15] Examples of conflict originating over administrative issues may involve disagreements over the project manager's authority and responsibilities, reporting relationships, administrative support, status reviews, or interorganizational interfacing. For the most part, disagreements over administrative procedures involve issues of how the project manager will function and how he relates to the organization's top management.

Personality conflict was ranked low in intensity by the project managers. Our discussions with project managers indicated that while the intensity of personality conflicts may not be as high as some of the other sources of conflict, they are among the most difficult to deal with effectively. Personality issues also may be obscured by communication problems and technical issues. A support person, for example, may stress the technical aspect of a disagreement with the project manager when, in fact, the real issue is a personality conflict.

Cost, like schedules, is often a basic performance measure in project management. As a conflict source, cost ranked lowest. Disagreements

[15] See Thamhain and Wilemon in footnote 11.

over cost frequently develop when project managers negotiate with other departments who will perform subtasks on the project. Project managers with tight budget constraints often want to minimize cost, while support groups may want to maximize their part of the project budget. In addition, conflicts may occur as a result of technical problems or schedule slippages which may increase costs.

Conflict Sources and Intensity in the Project Life Cycle

While it is important to examine some of the principal determinants of conflict from an aggregate perspective, more specific and useful insights can be gained by exploring the intensity of various conflict sources in each life-cycle stage, namely, project formation, project build-up, main program phase, and phaseout. See Figure 6-2.

1. *Project formation:* As Figure 6-2 illustrates, during the project formation state, the following conflict sources listed in order of rank were found:[16]

1. Project priorities
2. Administrative procedures
3. Schedules
4. Manpower
5. Cost
6. Technical
7. Personality

Unique to the project formation phase are some characteristics not typical of the other life-cycle stages. The project manager, for example, must launch his project within the larger "host" organization. Frequently, conflict develops between the priorities established for the project and the priorities that other line and staff groups believe important. To eliminate or minimize the detrimental consequences that could result, project managers need to carefully evaluate and plan for the impact of their projects on the groups that support them. This should be accomplished as early as possible in the project life cycle. The source of conflict ranked second

[16] Conflict intensity is computed as the total frequency (F) × magnitude (M) product of conflict experienced within the sample of project managers, when $0 \leq M \leq 3$. For example, if the average conflict intensity experienced by project managers on schedules with all interfaces was $M = 1.65$ (considerable) and $F = 14\%$ of all project managers indicated that "most" of this conflict occurred during the project formation phase, then "conflict over schedules" would be $M \times F = 1.65 \times 0.14 = 0.23$ during project formation.

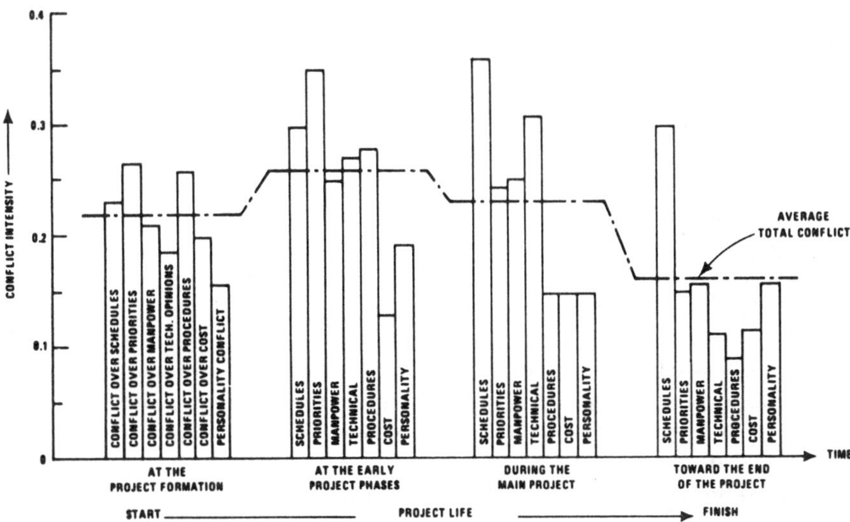

Figure 6-2. Relative intensity of conflict over the life cycle of projects. (*Source:* Hans J. Thamhain and David L. Wilemon, "Conflict Management in Project Life Cycles," *Sloan Management Review,* Summer 1975, pp. 31–50. Reprinted by permission.)

was administrative procedures, which are concerned with several critically important management issues; for example: How will the project organization be designed? Who will the project manager report to? What is the authority of the project manager? Does the project manager have control over manpower and material resources? What reporting and communication channels will be used? Who establishes schedules and performance specifications? Most of these areas are negotiated by the project manager, and conflict frequently occurs during the process. To avoid prolonged problems over these issues, it is important to clearly establish these procedures as early as possible.

Schedules typify another area where established groups may have to accommodate the newly formed project organization by adjusting their own operations. Most project managers attest that this adjustment is highly susceptible to conflict, even under ideal conditions, since it may involve a reorientation of present operating patterns and "local" priorities in support departments. These same departments might be fully committed to other projects. For similar reasons, negotiations over support personnel and other resources can be an important source of conflict in the project formation stage. Thus, effective planning and negotiation over these issues at the beginning of a project appear important.

2. *Project buildup:* The conflict sources for the project build-up are listed below in order of rank:

1. Project priorities
2. Schedules
3. Administrative procedures
4. Technical
5. Manpower
6. Personality
7. Cost

Disagreements over project priorities, schedules, and administrative procedures continue as important determinants of conflict. Some of these sources of conflict appear as an extension from the previous program phase. Additional conflicts surface during negotiations with other groups in the buildup phase. It is interesting to note that while schedules ranked third in conflict intensity in the project formation phase, they are the second major conflict determinant in the buildup phase. Many of the conflicts over schedules arise in the first phase because of the disagreements that develop over the establishment of schedules. By contrast, in the buildup phase, conflict may develop over the enforcement of schedules according to objectives of the overall project plan.

An important point is that conflict over administrative procedures becomes less intense in the buildup phase, indicating the diminishing magnitude and frequency of administrative problems. It also appears that it is important to resolve potential conflicts, such as administrative disagreements, in the earlier phase of a project to avoid a replication of the same problems in the more advanced project life-cycle phases.

Conflict over technical issues also becomes more pronounced in the buildup phase, rising from the sixth-ranked conflict source in the project formation phase to fourth in the buildup phase. Often this results from disagreements with a support group that cannot meet technical requirements or wants to enhance the technological input for which it is responsible. Such action can adversely affect the project manager's cost and schedule objectives.

Project managers emphasized that personality conflicts are particularly difficult to handle. Even apparently small and infrequent personality conflicts might be more disruptive and detrimental to overall project effectiveness than intense conflicts over nonpersonal issues, which can often be handled on a more rational basis. Many project managers also indicated that conflict over cost in the buildup phase generally tends to be low for two primary reasons. First, conflict over the establishment of cost targets does not appear to create intense conflicts for most project managers. Second, some projects are not yet mature enough in the buildup phase to cause disagreements over cost between the project manager and those who support him.

3. *Main program:* The main program phase revels a different conflict pattern. The seven potential causes of conflict are listed in rank order below:

1. Schedules
2. Technical
3. Manpower
4. Priorities
5. Procedures
6. Cost
7. Personality

In the main program phase, the meeting of schedule commitments by various support groups becomes critical to effective project performance. In complex task management, the interdependency of various support groups dealing with complex technology frequently gives rise to slippages in schedules. When several groups or organizations are involved, this in turn can cause a "whiplash" effect throughout the project. In other words, a slippage in schedule by one group may affect other groups if they are on the critical path of the project.

As noted, while conflicts over schedules often develop in the earlier project phases, they are frequently related to the establishment of schedules. In the main program phase, our discussions with project managers indicated that conflicts frequently develop over the "management and maintenance" of schedules. The latter, as indicated in Figure 6-2, produces more intense conflicts.

Technical conflicts are also one of the most important sources of conflict in the main program phase. There appear to be two principal reasons for the rather high level of conflict in this phase. First, the main program phase is often characterized by the integration of various project subsystems for the first time, such as configuration management. Owing to the complexities involved in this integration process, conflicts frequently develop over lack of subsystem integration and poor technical performance of one subsystem, which may, in turn, affect other components and subsystems. Second, the fact that a component can be designed in prototype does not always assure that all the technical anomalies will be eliminated. In extreme cases, the subsystem may not even be producible in the main program phase. Such problems can severely impact the project and generate intense conflicts. Disagreements also may arise in the main program phase over reliability and quality control standards, various design problems, and testing procedures. All these problems can severely impact the project and cause intense conflicts for the project manager.

Manpower resources ranked third as a determinant of conflict. The need for manpower reaches the highest levels in the main program phase. If support groups also are providing personnel to other projects, severe strains over manpower availability and project requirements frequently develop.

Conflict over priorities continued its decline in importance as a principal cause of conflict. Again, project priorities tend to be a form of conflict most likely to occur in the earlier project phases. Finally, administrative procedures, cost, and personality were about equal as the lowest-ranked conflict sources.

4. *Phaseout:* The final stage, project phaseout, illustrates an interesting shift in the principal cause of conflict. The ranking of the conflict sources in this final project phase are

1. Schedules
2. Personality
3. Manpower
4. Priorities
5. Cost
6. Technical
7. Procedures

Schedules are again the most likely form of conflict to develop in project phaseout. Project managers frequently indicated that many of the schedule slippages that developed in the main program phase tended to carry over to project phaseout. Schedule slippages often become cumulative and impact the project most severely in the final stage of a project.

Somewhat surprisingly, personality conflict was the second-ranked source of conflict. It appears that much of the personality-oriented conflict can be explained in two ways. First, it is not uncommon for project participants to be tense and concerned with future assignments. Second, project managers frequently note that interpersonal relationships may be quite strained during this period owing to the pressure on project participants to meet stringent schedules, budgets, and performance specifications and objectives.

Somewhat related to the personality issue are the conflicts that arise over manpower resources, the third-ranked conflict source. Disagreements over manpower resources may develop due to new projects phasing in, hence creating competition for personnel during the critical phaseout stage. Project managers, by contrast, also may experience conflicts over the absorption of surplus manpower back into the functional areas where they impact the budgets and organizational variables.

Conflict over priorities in the phaseout stage often appears to be directly or indirectly related to competition with other project start-ups in the organization. Typically, newly organized projects or marketing support activities might require urgent, short-notice attention and commitments that have to be squeezed into tight schedules. At the same time, personnel might leave the project organization prematurely because of prior commitments that conflict with a slipped schedule on the current project or because of a sudden opportunity for a new assignment elsewhere. In either case, the combined pressure on schedules, manpower, and personality creates a climate that is highly vulnerable to conflicts over priorities.

As noted in Figure 6-2, cost, technical, and administrative procedures tend to be ranked lowest as conflict sources. Cost, somewhat surprisingly, was not a major determinant of conflict. Discussions with project personnel suggest that while cost control can be troublesome in this phase, intense conflicts usually do not develop. Most problems in this area develop gradually and provide little ground for arguments.[17] The reader should be cautioned, however, that the low level of conflict is by no means indicative of the importance of cost performance to overall rating of a project manager. During discussions with top management, it was repeatedly emphasized that cost performance is one of the key evaluation measures in judging the performance of project managers.

Technical and administrative procedures ranked lowest in project phaseout. When a project reaches this stage, most of the technical issues are usually resolved. A similar argument holds for administrative procedures.

A graphical summary of the relative conflict intensity over the four conflict stages is provided in Figure 6-3. The diagram, an abstract of

[17] Depending on the work environment and particular business, there might be various reasons why conflicts over cost are low. First, some of the project components may be purchased externally on a fixed-fee basis. In such cases the contractor would bear the burden of costs. Second, costs are one of the most difficult project variables to control throughout the life cycle of a project, and budgets are frequently adjusted for increase in material and manpower costs over the life of the project. These incremental cost adjustments frequently eliminate some of the "sting" in cost when they exceed the original estimates of the project manager. Moreover, some projects in the high-technology area are managed on a cost-plus basis. In some of these projects precise cost estimates cannot always be rigidly adhered to.

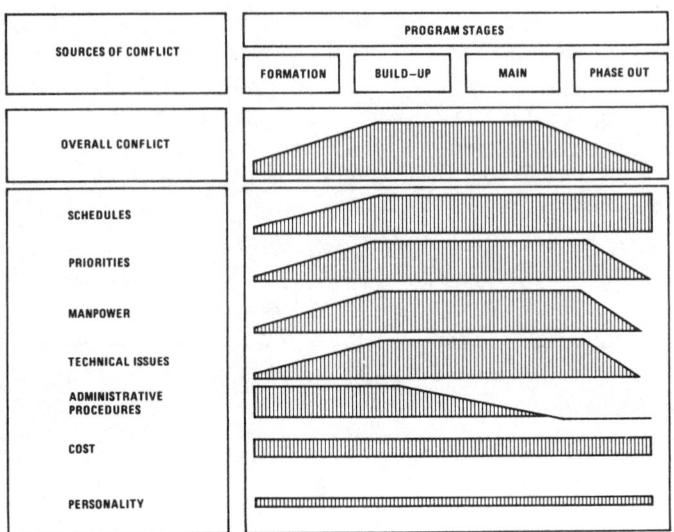

Figure 6-3. Trend of conflict intensity over the four project life-cycle stages. (*Source:* Hans J. Thamhain and David L. Wilemon, "Conflict Management in Project Life Cycles," *Sloan Management Review*, Spring 1975, pp. 31–50. Reprinted by permission.)

Figures 6-1 and 6-2, shows the change of relative conflict intensity over the project life for each of the seven conflict sources.

It is important to note that while a determinant conflict may be ranked relatively low in a specific life-cycle stage, it can, nevertheless, cause severe problems. A project manager, for example, may have serious ongoing problems with schedules throughout his project, but a single conflict over a technical issue can be equally detrimental and could jeopardize his performance to the same extent as schedule slippages. This point should be kept in mind in any discussion on project management conflict. Moreover, problems may develop that are virtually "conflict-free" (i.e., technological anomalies or problems with suppliers) but may be just as troublesome to the project manager as any of the conflict issues discussed.

The problem that now should be addressed is how these various conflict sources and the situations they create are managed.

Conflict-handling Modes

The investigation into the conflict-handling modes of project managers developed a number of interesting patterns. The actual style of project managers was determined from their scores on the aphorisms. As indi-

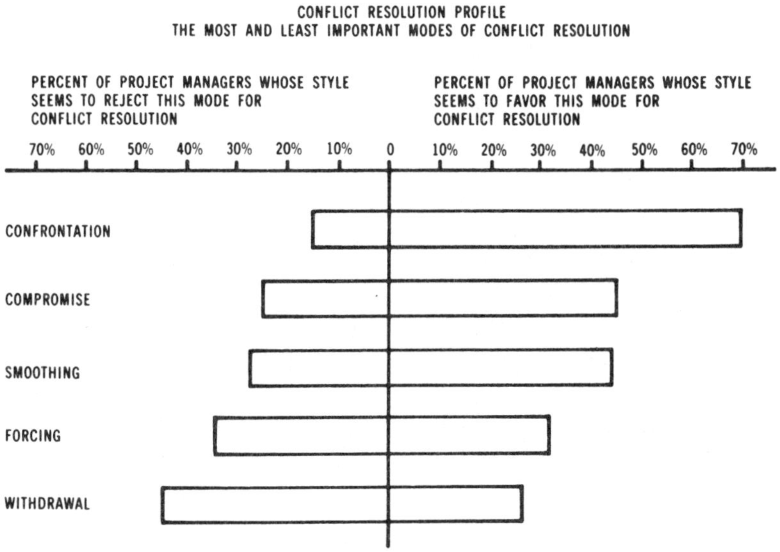

CONFLICT RESOLUTION PROFILE
THE MOST AND LEAST IMPORTANT MODES OF CONFLICT RESOLUTION

PERCENT OF PROJECT MANAGERS WHOSE STYLE SEEMS TO REJECT THIS MODE FOR CONFLICT RESOLUTION

PERCENT OF PROJECT MANAGERS WHOSE STYLE SEEMS TO FAVOR THIS MODE FOR CONFLICT RESOLUTION

70% 60% 50% 40% 30% 20% 10% 0 10% 20% 30% 40% 50% 60% 70%

CONFRONTATION

COMPROMISE

SMOOTHING

FORCING

WITHDRAWAL

Figure 6-4. Conflict resolution profile. The various modes of conflict resolution actually used to manage conflict in project-oriented work environments. (*Source:* Hans J. Thamhain and David L. Wilemon, "Conflict in Project Life Cycles," *Sloan Management Review*, Summer 1975, pp. 31–50. Reprinted by permission.)

cated in Figure 6-4, confrontation was most frequently utilized as a problem-solving mode. This mode was favored by approximately 70% of the project managers.[18] The compromise approach which is characterized by trade-offs and a give-and-take attitude ranked second, followed by smoothing. Forcing and withdrawal ranked as the fourth and fifth most favored resolution modes, respectively.

In terms of the most and least favored conflict resolution modes, the project managers had similar rankings for the conflict-handling method used between him and his personnel, his superior, and his functional support departments except in the cases of confrontation and compromise. While confrontation was the most favored mode for dealing with superiors, compromise was more favored in handling disagreements with functional support departments. The various modes of conflict resolution used by project managers are summarized in the profile of Figure 6-4.

[18] Quantitatively, this means that 70% of the project managers in the sample indicated that the proverbs that are representative of this mode (i.e., confrontation) describe the actual way the manager is resolving conflict "accurately" or "very accurately" as related to his project situations.

Implications

A number of ideas evolved for improving conflict management effectiveness in project-oriented environments from this research. As the data on the mean conflict intensities indicate, the three areas most likely to cause problems for the project manager over the entire project cycle are disagreements over schedules, project priorities, and manpower resources. One reason these areas are apt to produce more intense disagreements is that the project manager may have limited control over other areas that have an important impact on these areas particularly the functional support departments. These three new (schedules, project priorities, and manpower resources) require careful surveillance throughout the life cycle of a project. To minimize detrimental conflict, intensive planning prior to actually launching of the project is recommended. Planning can help the project manager anticipate many potential sources of conflict before they occur. Scheduling, priority setting, and resource allocation require effective planning to avoid problems later in the project. In our discussions with project managers who have experienced problems in these areas, almost all maintain that these problems frequently originate from lack of effective pre-project planning.

Managing projects involves managing change. It is not our intention to suggest that all such problems can be eliminated by effective planning. A more realistic view is that many potential problems can be minimized. There always will be random, unpredictable situations that defy forecasting in project environments.

Some specific suggestions are summarized in Table 6-1. The table provides an aid to project managers in recognizing some of the most important sources of conflict that are most likely to occur in various phases of projects. The table also suggests strategies for minimizing their detrimental consequences.

As one views the seven potential sources of disagreements over the life of a project, the dynamic nature of each conflict source is revealed. Frequently, areas that are most likely to foster disagreements early in a project become less likely to induce severe conflicts in the maturation of a project. Administrative procedures, for example, continually lose importance as an intense source of conflict during project maturation. By contrast, personality conflict, which ranks lowest in the project formation stage, is the second most important source of conflict in project phaseout. In summary, it is posited that if project managers are aware of the importance of each potential conflict source by project life cycle, then more effective conflict minimization strategies can be developed.

Table 6-1. Major Conflict Source and Recommendations for Minimizing Dysfunctional Consequences.

PROJECT LIFE CYCLE PHASE	CONFLICT SOURCE	RECOMMENDATIONS
Project formation	Priorities	Clearly defined plans. Joint decision-making and/or consultation with affected parties.
	Procedures	Develop detailed administrative operating procedures to be followed in conduct of project. Secure approval from key administrators. Develop statement of understanding or charter.
	Schedules	Develop schedule commitments in advance of actual project commencement. Forecast other departmental priorities and possible impact on project.
Buildup phase	Priorities	Provide effective feedback to support areas on forecasted project plans and needs via status review sessions.
	Schedules	Schedule work breakdown packages (project subunits) in cooperation with functional groups.
	Procedures	Contingency planning on key administrative issues.
Main program	Schedules	Continually monitor work in progress. Communicate results to affected parties. Forecast problems and consider alternatives. Identify potential "trouble spots" needing closer surveillance.
	Technical	Early resolution of technical problems. Communication of schedule and budget restraints to technical personnel. Emphasize adequate, early technical testing. Facilitate early agreement on final designs.
	Manpower	Forecast and communicate manpower requirements early.
	Manpower	Establish manpower requirements and priorities with functional and staff groups.
Phaseout	Schedules	Close schedule monitoring in project life cycle. Consider reallocation of available manpower to critical project areas prone to schedule slippages. Attain prompt resolution of technical issues which may impact schedules.
	Personality and Manpower	Develop plans for reallocation of manpower upon project completion. Maintain harmonious working relationships with project team and support groups. Try to loosen up "high-stress" environment.

Source: Hans J. Thamhain and David L. Wilemon, "Conflict Management in Project Life Cycles," *Sloan Management Review*, Summer 1975, pp. 31–50. Reprinted by permission.

In terms of the means by which project managers handle conflicts and disagreements, the data revealed that the confrontation or problem-solving mode was the most frequent method utilized. While our study did not attempt to explore the effectiveness of each mode separately, in an earlier research project Burke[19] suggests that the confrontation approach is the most effective conflict-handling mode.[20]

In some contrast to studies of general management, the findings of our research in project-oriented environments suggest that it is less important to search for a best mode of effective conflict management. It appears to be more significant that project managers, in their capacity as integrators of diverse organizational resources, employ the full range of conflict resolution modes. While confrontation was found as the ideal approach under most circumstances, other approaches may be equally effective depending upon the situational content of the disagreement. Withdrawal, for example, may be used effectively as a temporary measure until new information can be sought, or to "cool off" a hostile reaction from a colleague. As a basic long-term strategy, however, withdrawal may actually escalate a disagreement if no resolution is eventually sought.[21]

In other cases, compromise and smoothing might be considered an effective strategy by the project manager, if it does not severely affect the overall project objectives. Forcing, on the other hand, often proves to be a win/lose mode. Even though the project manager may win over a specific issue, effective working arrangements with the "forced" party may be jeopardized in future relationships. Nevertheless, some project managers find that forcing is the only viable mode in some situations. Confrontation of the problem-solving mode may actually encompass all conflict-handling modes to some extent. A project manager, for example, in solving a conflict may use withdrawal, compromise, forcing, and smoothing to eventually get an effective resolution. The objective of confrontation, however, is to find a solution to the issue in question whereby all affected parties can live with the eventual outcome.

In summary, conflict is fundamental to complex task management. It is important not only for project managers to be cognizant of the potential

[19] R. J. Burke, "Methods of Resolving Interpersonal Conflict," *Personnel Administration*, July–August 1969, pp. 48–55.

[20] Although Burke's study was conducted on general management personnel, it offers an interesting comparison to our research. Burke's paper notes that "Compromising" and "Forcing" were effective in 11.3% and 24.5% of the cases, while "Withdrawal" or "Smoothing" approaches were found mostly ineffective in the environment under investigation.

[21] H. J. Thamhain and D. L. Wilemon, "Conflict Management in Project-Oriented Work Environments," *Proceedings of the Sixth International Meeting of the Project Management Institute*, Washington, D.C., September 1974, pp. 18–21.

sources of conflict, but also to know when in the life cycle of a project they are most likely to occur. Such knowledge can help the project manager avoid the detrimental aspects of conflict and maximize its beneficial aspects. Conflict can be beneficial when disagreements result in the development of new information that can enhance the decision-making process. Finally, when conflicts do develop, the project manager needs to know the advantages and disadvantages of each resolution mode for conflict resolution effectiveness.

CONFLICT RESOLUTION

Although each project within the company may be inherently different, the company may wish to have the resulting conflicts resolved in the same manner. The four most common methods are

1. The development of company-wide conflict resolution policies and procedures
2. The establishment of project conflict resolutions procedures during the early planning activities
3. The use of hierarchical referral
4. The requirement of direct contact

With each of the above methods, the project manager may still select any of the conflict resolution modes discussed in the previous section.

Many companies have attempted to develop company-wide policies and procedures for conflict resolution. Results have shown that this method must be judiciously used because each project is different and not all conflicts can be handled the same way. Furthermore, project managers, by virtue of their individuality, and sometimes differing amounts of authority and responsibility, prefer to resolve conflicts in their own fashion.

Cleland and King have suggested a generalized model for effecting the resolution of conflict which in practice reduces the need for resolution of the conflict by the common line supervisor. They suggest that an organizational policy or procedure be issued which encourages the resolution of the conflict on the following:

1. Each project and functional manager will assume the initiative in taking actions which relate to his or her particular responsibilities and authorities. It should be the policy of the organization to emphasize the resolution of these disagreements through negotiation below the general-manager level. If, and only if, mutual resolution is not

found quickly through such negotiation, and if the issue is a salient project-functional one, the project and functional managers should seek an audience with their mutual superior for resolution of the issue.

2. Before such an audience is given, each manager must clearly try to reach agreement on the following:
   a. *The issue*
   b. *The impact* upon the project and the total organization
   c. *The alternatives,* examined together with the related costs and benefits of each
   d. *The recommendation* (as each sees it)
3. Further, each participant should be fully aware of the issues, their impact, and the alternatives and recommendations offered by the other. If this procedure is followed, it should reduce the opportunity for *personal* conflict and keep the conflict *organizational* and minimal in nature.[22]

A second method for resolving conflicts, and one that is often very effective, is to "plan" for conflicts during the planning activities. This can be accomplished through the use of linear responsibility charts. Planning for conflict resolution is similar to the first method except that each project manager can develop his own policies, rules, and procedures.

Hierarchical referral for conflict resolution, in theory, appears as the best method because neither the project manager nor the functional manager will dominate. Under this arrangement, the project and functional managers agree that for a proper balance to exist. Their common superior must resolve the conflict in order to protect the company's best interest. Unfortunately, this is not a realistic course of action because the common superior cannot be expected to continually resolve lower-level conflicts. Going to the well too often gives the impression that the functional and project managers cannot resolve their own problems.

The last method is direct contact, which is an outgrowth of the policies and procedures methods where established guidelines dictate that conflicting parties meet face to face and resolve their disagreement. Unfortunately, this method does not always work and, if continually stressed, can result in conditions where individuals will either suppress the identification of problems or develop new ones during confrontation.

Many conflicts can be either reduced or eliminated by constant communication of the project objectives to the team members. Many times

[22] David I. Cleland and William R. King, *Systems Analysis and Project Management,* 3rd ed. (New York: McGraw-Hill, 1983), p. 344.

this continual repetition will prevent individuals from going too far into the "wrong" and thus avoid the creation of a conflict situation.

UNDERSTANDING SUPERIOR, SUBORDINATE, AND FUNCTIONAL CONFLICTS[23]

In order for the project manager to be effective, an understanding of how to work with the various employees who must interface with the project is necessary. These various employees include upper-level management, subordinate project team members, and functional personnel. Quite often, especially when conflicts are possible, the project manager must demonstrate an ability for continuous adaptability by creating a different working environment with each group of employees. The need for this was shown in the previous section by the fact that the relative intensity of conflicts can vary in the life cycle of a project.

The type and intensity of conflicts can also vary with the type of employee whom the project manager must interface with, as shown in Figure 6-5. Both conflict causes and sources are rated according to relative conflict intensity. Any conflict that the project manager has with a functional manager can also occur with the functional employee, and vice versa. The data in Figure 6-5 were obtained for a 75% confidence level.

In the previous section we discussed the five basic resolution modes for handling conflicts. The specific type of resolution mode that a project manager will use might easily depend upon whom the conflict is with, as shown in Figure 6-6. The data in Figure 6-6 do not necessarily show the modes that project managers would prefer, but rather identify the modes that will increase or decrease the potential conflict intensity. For example, although project managers consider, in general, that withdrawal is their least favorite mode, it can be used quite effectively with functional managers. In dealing with superiors, project managers would rather be ready for an immediate compromise than for face-to-face confrontation which could easily result in having the resolution forced in the favor of upper-level management.

Figure 6-7 identifies the various influence styles that project managers find effective in helping to reduce potential conflicts. Penalty power, authority, and expertise are considered as strongly unfavorable associations with respect to low conflicts. As expected, work challenge and promotions (if the project manager has the authority) are strongly favorable associations with his personnel.

[23] The majority of this section, including the figures, has been adapted from *Seminar in Project Management Workbook*, © 1977 by Hans J. Thamhain. Reproduced by permission of Dr. Hans J. Thamhain.

CONFLICT CAUSES	SOURCES: CONFLICTS OCCURRED MOSTLY WITH				
	FUNCTIONAL MANAGERS	FUNCTIONAL PERSONNEL	BETWEEN PROJECT PERSONNEL	SUPERIORS	SUBORDINATES
SCHEDULES	■	■			
PRIORITIES	■	■	■		
MANPOWER	■	■			
TECHNICAL	■	■	■		
PROCEDURES	■	■		■	■
PERSONALITY	■	■	■	■	■
COSTS	■	■		■	

HIGH ← RELATIVE CONFLICT INTENSITY → LOW (across columns)

LOW ← RELATIVE CONFLICT INTENSITY → HIGH (up the rows)

Figure 6-5. Relationship between conflict causes and sources.

INTENSITY OF CONFLICT PERCEIVED BY PROJECT MANAGERS (P.M.)	ACTUAL CONFLICT RESOLUTION STYLE				
	FORCING	CONFRONTATION	COMPROMISE	SMOOTHING	WITHDRAWAL
BETWEEN P.M. AND HIS PERSONNEL	■	▲	▲	▲	■
BETWEEN P.M. AND HIS SUPERIOR		■	▲		
BETWEEN P.M. AND FUNCTIONAL SUPPORT DEPARTMENTS	■	■			▲

▲ STRONGLY FAVORABLE ASSOCIATION WITH REGARD TO LOW CONFLICT (– τ)

■ STRONGLY UNFAVORABLE ASSOCIATION WITH REGARD TO LOW CONFLICT (+ τ)

*KENDALL τ CORRELATION

Figure 6-6. Association between perceived intensity of conflict and mode of conflict resolution.* (The figure shows only those associations which are statistically significant at the 95%.)

INTENSITY OF CONFLICT PERCEIVED BY PROJECT MANAGER (P.M.)	INFLUENCE METHODS AS PERCEIVED BY PROJECT MANAGERS						
	EXPERTISE	AUTHORITY	WORK CHALLENGE	FRIENDSHIP	PROMOTION	SALARY	PENALTY
BETWEEN P.M. AND HIS PERSONNEL	■	■	▲		▲		■
BETWEEN P.M. AND HIS SUPERIOR			▲				■
BETWEEN P.M. AND FUNCTIONAL SUPPORT DEPARTMENTS		■					■

▲ STRONGLY FAVORABLE ASSOCIATION WITH REGARD TO LOW CONFLICT (– τ)

■ STRONGLY UNFAVORABLE ASSOCIATION WITH REGARD TO LOW CONFLICT (+ τ)

*KENDALL τ CORRELATION

Figure 6-7. Association between influence methods of project manager and their perceived conflict intensity.* (The figure shows only those associations which are statistically significant at the 95% level.)

Therefore, for the project manager to be truly effective, he or she should understand not only what types of conflicts are possible in the various stages of the life cycle, but with whom these conflicts can occur and how to deal with them effectively.

Good project managers realize that conflicts are inevitable and that procedures or techniques must be developed for their resolution. If the project manager is not careful, he or she could easily worsen the conflict by not knowing how to manage it. Once a conflict occurs, the project manager must observe certain preliminaries, including

- Studying the problem and collecting all available information
- Developing a situational approach or methodology
- Setting the appropriate atmosphere or climate

In setting the appropriate atmosphere, the project manager must establish a willingness to participate for himself as well as the other participants. The manager must clearly state the objectives of the forthcoming meeting, establish the credibility of the meeting, and sanction the meeting.

If a confrontation meeting is necessary between conflicting parties, then the project manager should be aware of the logical steps and sequence of events that should be taken. These include the following:[24]

[24] See footnote 23.

- Setting the climate: establishing a willingness to participate
- Analyzing the images (How do you see yourself and others, and how do they see you?)
- Collecting the information: getting feelings out in the open
- Defining the problem: defining and clarifying all positions
- Sharing the information: making the information available to all
- Setting the appropriate priorities: developing working sessions for setting priorities and time tables
- Organizing the group: forming cross-functional problem-solving groups
- Problem solving: obtaining cross-functional involvement, securing commitments, and setting the priorities and time table
- Developing the action plan: getting commitment
- Implementing the work: taking action on the plan
- Following up: obtaining feedback on the implementation for the action plan

Once the conflict has been defined and a meeting is necessary, the project manager or team leader should understand the conflict minimization procedures. These include

- Pausing and thinking before reacting
- Building trust
- Trying to understand the conflict motives
- Keeping the meeting under control
- Listening to all involved parties
- Maintaining a give-and-take attitude
- Educating others tactfully on your views
- Being willing to say when you were wrong
- Not acting as a superman and leveling the discussion only once in a while

We can now sum up these actions by defining the role of the effective manager in conflict problem solving. The effective manager

- Knows the organization
- Listens with understanding rather than evaluation
- Clarifies the nature of the conflict
- Understands the feelings of others
- Suggests the procedures for resolving differences

- Maintains relationships with disputing parties
- Facilitates the communications process
- Seeks resolutions

SUMMARY

Conflict is endemic in team management and should be dealt with in a forthright manner by both managers and professionals. Conflict arises for many reasons and can occur with anyone and over anything. Conflict can be managed; conflict emerges at every phase of the life cycle of projects dealt with by engineering teams.

Different techniques can be developed for dealing with conflict. The ability to communicate is important to any mode of conflict resolution. Above all, managers and professionals should recognize that conflict in teams is normal; interpersonal conflict which deteriorates into personal animosities and attacks should be avoided. When conflicts arise, all persons concerned should be encouraged to "talk out" the conflict and seek a resolution that preserves the integrity of the team and the dignity of the team members.

Decision making is the topic discussed in Chapter 7.

BIBLIOGRAPHY

Blake, R. R., and J. S. Mouton. *The Managerial Grid.* Houston: Gulf Publishing, 1974.

Burke, R. J. "Methods of Resolving Interpersonal Conflict." *Personnel Administration,* July–August 1969.

Burke, R. J. "Methods of Managing Superior-Subordinate Conflict." *Canadian Journal of Behavioral Science,* vol. 2, no. 2, 1970.

Cleland, David I., and William R. King. *Systems Analysis and Project Management,* 3rd ed. New York: McGraw-Hill, 1983.

"Diagnosing Conflict Determinants in Project Management." *IEEE Transactions on Engineering Management,* vol. 22, 1975.

Dugan, H. S., H. J. Thamhain, and D. L. Wilemon. "Managing Change in Project Management." *Proceedings of the Ninth Annual International Seminar/Symposium on Project Management,* Chicago, October 22–26, 1977.

Lawrence, P. R., and J. W. Lorsch. "New Management Job: the Integrator." *Harvard Business Review,* November–December 1967.

Thamhain, H. J., and D. L. Wilemon. "Conflict Management in Project-Oriented Work Environments." *Proceedings of the Sixth International Meeting of the Project Management Institute,* Washington, D.C., September 18–21, 1974.

Thamhain, Hans J., and David L. Wilemon. "Conflict Management in Project Life Cycles." *Sloan Management Review,* Summer 1975.

Wilemon, D. L., "Project Management Conflict: A View From Apollo." *Proceedings of the Third Annual Symposium of the Project Management Institute,* Houston, Texas, October 1971.

Wilemon, David L. "Managing Conflict in Temporary Management Situations." *The Journal of Management Studies,* 1973.

Wilemon, D. L., and J. P. Cicero. "The Project Manager—Anomalies and Ambiguities." *Academy of Management Journal,* Fall 1970.

7
DECISION MAKING

There are two key activities carried out in the management of a team: deciding what needs to be done, and then doing it. In this chapter we deal with the decision process in managing a team.

A decision is the act or process of selecting a course of action after consideration of the alternative ways that resources can be used to attain organization purposes. In the process of making a decision, the decision maker—a team member or a professional on the team—carries out certain thought patterns:

- Evaluation of the current environment and situation
- Assessment of what the situation and environment will be when the decision will be implemented
- Analysis of how the decision will affect the "stakeholders" who have an interest in the outcome of the decision
- Selection of a strategy on how the decision will be implemented

Decision making is the process through which alternatives are evaluated and a course of action is selected as the solution to a problem, opportunity, or issue.

Team decisions are inseparable from the team and the organizational planning elements: mission, objectives, goals, and strategies. Every principal decision made in the context of a team's activities should help the team members to decide

- What is to be done?
- How will it be done?
- Who will do it?
- Why is it being done?
- When is it to be done?

It is through the decision process that a determination of future action for the team is made. The decision-making process is a rational attempt to achieve results. Decision making encompasses both an *active* and *passive* use of resources.

An *active* decision is one made through full consideration of existing mission, objectives, goals, and strategies, and a sincere attempt to make effective and efficient use of resources. A *passive* decision also consumes resources. If a decision maker is not active in making a decision and procrastinates, organizational resources are consumed during the period of procrastination. If no deliberate decision is made, in reality a decision has been made to do nothing. Resources continue to be consumed in the same manner that has been done in the past, often to the neglect of stakeholder consideration.

Stakeholders are the persons, organizations, or institutions who contribute to or receive benefit from the decision. A product design team would have stakeholders composed of manufacturing, reliability, quality control, and design engineers plus representatives from customer service, marketing, and production control. Each of these professionals, as well as the organizations they represent, would have a vested and rightful interest in the decisions that are made in the design of the product. A decision made by the team leader on behalf of the team should be worked through the stakeholders. There is a practical reason for doing this.

A decision has to be implemented through a process of

- Delegation of appropriate authority and responsibility
- Allocation of human and nonhuman resources to support the decision
- Commitment of the people to make the decision workable
- Design of a monitoring and control system to determine if the decision is accomplishing the results that are desired

Key decisions in modern organizations are influenced by laws, government regulations, environmentalists, and the actions or reactions of

- Competitors
- Suppliers
- Customers
- Collective bargaining units
- Employees
- Stockholders
- Creditors
- Local communities
- Professional societies

Consequently, the decision maker should always be mindful of how key decisions will be received by stakeholders who have a vested interest in the outcome of the decision to resolve a problem or opportunity.

PROBLEMS AND OPPORTUNITIES

A problem is a question raised for inquiry, consideration, or solution. Failure to meet product quality standards during manufacture is a problem. Inventory shortages during a production run are a problem. These problems, if viewed positively, can become an opportunity for the astute manager.

An opportunity is a favorable chance for advancement or progress. If a company's manufacturing process fails to produce a quality product, then the opportunity exists to improve product quality through

- Review of product design
- Analysis of manufacturing engineering standards
- Assessment of quality inspection techniques
- Evaluation of adequacy of manufacturing management
- Investigation of motivation and commitment of manufacturing personnel to implement adequate quality policy and procedures

The difference between a problem and an opportunity depends on the beholder. However, problems and opportunities should be differentiated. David B. Gleicher, a management consultant, distinguishes between the two terms in the following way: A problem is "something that endangers the organization's ability to reach its objectives, while an opportunity is something that offers the chance to exceed objectives."[1]

Peter Drucker makes it clear that opportunities rather than problems are the keys to organizational and managerial success. He observes that solving a problem merely restores normality; but results must come from the exploitation of opportunities. He links exploitation of opportunities to finding the right things to do, and concentrating "resources and efforts on them."[2]

Identifying problems or opportunities is a key activity of the manager. Successful managers do not wait for someone else to tell them what to do; they must figure that out for themselves.

William Pounds has described several situations that can alert managers to possible problems (issues or opportunities). *First,* when there is a deviation from past experience; *second,* when there is a deviation from an existing plan; *third,* when problems are presented to the manager by other

---

[1] Cited in James A. F. Stoner, *Management,* 2nd ed. (Englewood Cliffs, N.J.: Prentice-Hall, 1982), pp. 166–167.

[2] Peter F. Drucker, *Managing for Results* (New York: Harper & Row, 1964), p. 5.

individuals; and fourth, when competitors out-perform the manager's organization.[3]

THE FRAMEWORK OF DECISION MAKING

There are key considerations in a conceptual framework for decision making. These considerations are discussed below.

Judgmental Decisions

One type of decision is that based upon past experiences or knowledge, that is, a judgmental decision. The decision maker remembers what has happened in similar past situations and uses this data base to predict the outcome of the alternatives that are involved in the decision. This is a useful approach to decision making since many situations, problems, and opportunities do tend to recur frequently in organizations. A judgmental decision is faster and more inexpensive to make; however, it does have a drawback. It is based on common sense and memory, and some people do not have common sense or good memory. Then, too, judgmental decisions are not able to deal with those situations which are new and in which there is no experience base on which to select a logical course of action.

Rational Decisions

A rational decision is one in which a rational process is reasoned through to make the decision. Certain conditions must exist before we can say that people are approaching decision making in a rational fashion. In the *first* place, there must be some objectives and goals toward which resources can be committed. *Second,* the decision maker must develop a clear understanding of the alternative actions that can be taken to reach the objectives and goals. *Third,* they must have the information to analyze the alternatives in light of the goals or objectives that are being sought. *Finally,* the decision maker must have a desire to come to the best possible solution by selecting the alternative which best satisfies the overall effectiveness of the organization.

The reality of decision making, theoretical versus the real world, is described by Hampton in the following way: ". . . the ideal version of integrated, logical, goal-seeking organizations and rational decisions makers is a useful fiction, much like heroic figures are in the literature.

[3] William F. Pounds, "The Process of Problem Finding." See also Peter F. Drucker, *The Practice of Management* (New York: Harper & Row, 1954), pp. 351–354.

Heroes model virtues to aspire to. Ideal versions of managerial decision processes remind us of the possibilities of being more systematic and logical."[4]

It is difficult to find complete rationality in the decision process. In the first place, decisions are made today but impact the future, and the future almost invariably involves risks and uncertainties, some of which will not be known. Second, the alternatives that have to be analyzed to make a decision may never all be recognized, simply because there are so many different ways in which organizational purposes can be achieved. Finally, not all alternatives can be analyzed even with the most modern analytical techniques, simply because there are so many alternative ways of doing things and the decision maker is always faced with a time frame for the decision.

Certainty/Uncertainty

Certain decision making is done under conditions of certainty, although such conditions are rare. Decision making under certainty occurs when the manager knows exactly which environmental or competitive action will occur. That is, the manager is able to make perfectly accurate decisions on a routine basis. Decision making under certainty requires information that is nearly perfect. Unfortunately, such perfect information is difficult to obtain since some of the variables in a decision, such as how people will react to the decision, is never predictable with complete accuracy.

Three words can be used to describe the different degree of predictability in decision making. Professor Stoner has used these words to describe the different situations: "certainty," "risk," and "uncertainty." "Under conditions of *certainty,* we know what will happen in the future. Under *risk* we know what the probability of each possible outcome is. Under *uncertainty,* we do not know the probabilities—and maybe not even the possible outcomes."[5]

Controllable/Noncontrollable Factors

Managers operate within environments in which some of the decision variables are controllable and others are non-controllable. In a controlla-

---

[4] David R. Hampton, *Contemporary Management* (New York: McGraw-Hill, 1981), p. 221.

[5] See F. H. Knight, *Risk, Uncertainty, and Profit* (New York: Harper & Row, 1920); and James A. F. Stoner, *Management,* 2nd ed. (Englewood Cliffs, N.J.: Prentice-Hall, 1982), p. 163.

ble environment the decision maker can select a strategy and control the execution of that strategy.

In the decision making context the one major variable that is controllable is the actual choice of the strategy. However the implementation of that strategy will be affected by uncontrollable variables within the environment.

There are two variables which influence the selection of alternatives in a strategy, namely (1) the states of nature; and (2) competitive action. We use the term, "states of nature" to designate the possible events that may actually occur. For example, the uncontrollable state of nature facing all of us is death. For the business manager, competitive actions are uncontrollable.

States of nature are uncontrollable from the position of the manager. For example, the accuracy of forecasts will depend in a large major upon the state of the overall economy. In today's complex environment, there are a large number of possible states of nature that can occur. It's virtually impossible to list them all and to determine the effects that each might have on an outcome.

To illustrate this, consider the following example: Management has identified a niche in the marketplace and has placed you in charge of a new product development team. Your first requirement is to decide upon the size of the budget. Typical budgets range from $250,000 to $1,000,000. The larger the budget, the greater the results (i.e., objectives) expected by management.

Upper-level management, with the assistance of marketing, develop the following payoff table (matrix):

		STATES OF NATURE		
		HIGH DEMAND	MODERATE DEMAND	SMALL DEMAND
BUDGET	250,000	$1,000,000	750,000	500,000
SIZE	1,000,000	3,000,000	1,000,000	−500,000

If you select the small budget, profit payoffs will range from $500,000 to $1,000,000. If you select the large budget, some payoffs are large, but, based upon the state of nature, a loss of $500,000 can occur. This example states that there are 3 and only 3 states of nature when, in fact, there are many, many more.

The advantages of payoff matrices are the speed at which decision making can be accomplished and ease of performing contingency planning. The disadvantages are the difficulty in obtaining the payoff figures and the limitations in the number of states of nature.

Programmed/Nonprogrammed Decisions

Decisions can be classified as *programmed* or *nonprogrammed,* drawing on Professor Simon's classification.[6]

Programmed decisions are standing decisions that exist to guide managers and professionals in their daily organization life, existing organizational mission, objectives, goals, strategies, policies, standards, procedures, methods, and roles. Programmed decisions are guided by

1. *Policies,* which are guides set down by higher-level executives for the thinking of other managers and professionals in dealing with recurring anticipated situations. A product quality policy, a management development policy, and an inventory level policy are examples of policy. Discretion is expected in using policies as guidelines in programmed decisions.
2. *Standards,* which are criteria against which something can be compared. Engineering design standards, manufacturing output standards, and product quality standards are examples of criteria that guide decision making and implementation that are basically unchanging from day to day.
3. *Procedures,* which are sequential steps to carrying out action in the accomplishment of some tasks. Tasks are "procedurized" to reduce complexity, and ensure uniformity in their completion. Little discretionary behavior is permitted in using procedures. For example, the procedure for processing an engineering change on a construction project requires certain sequential actions that must be carried out to evaluate the cost, schedule, technical aspects, and contract implications of the change.
4. A *method,* which is one step of a procedure.[7] Methods provide the basis for an orderly methodology, development, or classification. A procedure for entering new orders on the company books would require specific methodology for entering the new order in the company's accounts receivable records.

Policies, standards, procedures, and methods provide a framework within which decisions can be made. Such a framework, reduces the

[6] See Herbert A. Simon, *The New Science of Management Decisions,* rev. ed. (Englewood Cliffs, N.J.: Prentice-Hall, 1977), p. 48.

[7] Stephen P. Robbins, *The Administrative Process: Integrating Theory and Practice* (Englewood Cliffs, N.J.: Prentice-Hall 1976), p. 157.

chance of error. It also reduces the time to make a decision. Prudent managers delegate authority and responsibility to make decisions and allocate resources within a framework of policy. Engineering design managers and professionals design the product within a product policy established by the senior executives of the company. A financial clerk processing an engineer's travel voucher uses established procedures and methodology to ascertain the validity of the travel costs, and sends on the voucher to the financial office for payment to the traveler.

The development by a manager of adequate policies, procedures, methods, and standards permits fuller delegation of decision-making authority and reduces the time the manager must actively supervise professionals. Then "management by exception" to the policies, procedures, standards, and methods can be carried out by the manager, conserving that individual's time.

If an organization has a well-defined strategic framework of mission, objectives, goals, and strategies, then key decisions regarding the allocation of resources can be made within it. This strategic framework becomes a baseline to guide key decision makers in the enterprise. In this sense, the strategic framework provides a basis for a "programmed" decision, within a framework of organizational mission, objectives, goals, and strategies.

Programmed decisions are those which are made in accordance with some preestablished guidelines, such as in a policy, procedure, rule, or methodology. If problems, opportunities, or issues recur, and if the integral elements can be defined, analyzed, and predicted, then we may develop supporting policies, procedures, methodologies, or rules in order that decision making can be "programmed." For example, the decisions on an inventory level to support a given product would involve fact finding and forecasting. If the analysis of the inventory level is properly performed, then from that a series of routine decision-making guidelines can be developed. Programmed decisions, however, do limit the freedom of the decision maker.

As an executive moves up the chain of command within an organization, the ability to deal with unprogrammed decisions becomes more important, since more decisions of an nonprogrammed nature are found at the top level of organizations.

Nonprogrammed decisions are unique and nonrecurring. Although the decision may be made within a strategic framework, the decision maker is faced with new alternatives and choices. The problem or opportunity is novel, inherently unstructured, usually involves unknown factors and forces. The selection of an enterprise's strategic framework, its mission, objectives, goals, and strategies are examples of nonprogrammed decisions. However, once established, this strategic framework provides the basis for certain programmed decisions to be made.

In nonprogrammed decisions, many different options are open to the manager. Creativity and innovation on the part of the decision maker are valued. Nonprogrammed decisions are found in every function of management: planning, organizing, motivating, leading, and controlling. The strategic planning function provides the greatest opportunity for nonprogrammed decisions since the decision maker is attempting to create something that does not currently exist. Intuition plays an important role in nonprogrammed decisions.

Some writers have defined the decision-making process as being primarily intuitive. Intuitive decisions are choices made solely on the basis of what a person feels is correct. The intuitive decision process plays a more important role in making decisions. Peter Schoderuek states, "While increasing amounts of information can considerably aid middle management in making decisions, the top echelon must still rely on intuitive judgment. Furthermore, . . . electronic computers do enable top management to focus more attention on data but they have not rendered obsolete the time-honored managerial intuitive know-how."[8]

Professor Mintzberg has verified the heavy reliance by top managers on intuition in the decision making process.[9]

Research into parapsychology has caused many people to acknowledge that certain individuals do have a remarkable aptitude for making a correct decision by intuition. There have been studies that indicate a correlation between decision success and possessing a high degree of extrasensory perception.[10] While the opportunity for intuitive decisions does exist, complex nonprogrammed decisions face monumental data for analysis, and a sizable number of variables. The alternatives are simply too complex and too numerous to rely on intuitive methods.

Professor Agor has described three broad types of management styles for decision making.[11] The first, using the left side of the brain, employs analytical, rational, and logical quantitative techniques, often using the computer as a "technological assistant." An alternative and complementary management style uses the right side of the brain. Feelings more than facts guide decision making; intuitive and inductive techniques are employed. Problems are solved by looking at the whole; participatory pro-

---

[8] Peter P. Schoderbek, *Management Systems,* 2nd ed. (New York: Wiley, 1971), p. 124.
[9] Henry Mintzberg, *The Nature of Managerial Work* (New York: Harper & Row, 1973).
[10] Jmihalasky and H. C. Sherwood, "Dollars May Flow from the Sixth Sense," *Nations Business,* **59,** (1971):64–66.
[11] Weston H. Agor, "Using Intuition to Manage Organizations in the Future," *Business Horizons,* July–August 1984, pp. 49–54.

cesses are used; decisions are made in an unstructured, fluid, and spontaneous manner. The third style (integrated) uses both left and right brain skills. In commenting on the integrated approach, Professor Agor notes:

> Managers who rely on this approach normally feel comfortable dealing with both facts and feelings when making decisions. But, they also tend to make their major decisions guided by intuition after scanning the available facts and receiving input from the management resources and personnel available both on the left and the right of the organization. Frequently, the intuitive decisions made are in conflict with the course suggested by the available facts and forward projections based thereon.[12]

The use of intuition is a critical skill in decision making, yet the subject has received little attention; research on the subject is extremely limited. However, recent articles have keynoted the importance of intuition in decision making.[13] Why is there a growing interest in intuition?

- There is a growing dissatisfaction with the left-brain style of decision making.
- Turbulent environmental changes—political, social, economic, and technological—are becoming a way of life. Extrapolation of the past is suspect.
- Employees as organizational "stakeholders" are demanding a greater role in organizational decision making.
- Managers are increasingly faced with management situations which are complex. Data bases will be limited and managers will have to rely on using both left and right brain skills.

Intuition will, according to Professor Agor, become more valuable in the making of management decisions in the future.[14]

How important is the decision? Managers make many decisions. Some are more important than others. Each manager should develop a set of guidelines to evaluate the importance of decisions and how much time and effort should be spent on a particular decision.

The importance of a decision depends in part on the organizational level at which the decision is being made. Decisions dealing with the establishment of corporate mission, objectives, goals, and strategies are of great

[12] Agor, op. cit., p. 50.
[13] Ibid.
[14] Agor, op. cit., p. 54.

importance. A decision on who works overtime on the production line is of less importance from a corporate viewpoint but of obvious importance to the individual who will be doing the actual overtime. Below are some guidelines that can help the decision maker to reach a judgment on the importance of the decision he or she is making.

1. *Size of length of the resource commitment that is being made:* A decision which commits resources over several years, such as new product development and design, is an important decision. A project to locate, design, and build a new plant requires detailed analysis.

2. *Impact on organizational "stakeholders":* If the decision will impact on several stakeholders, then care must be taken to factor in the probable impact, and acceptance, by the stakeholders.

3. *Impact on people:* If the careers and well-being of people are going to be affected by the decision, then great care should be taken to study the decision thoroughly.

4. *"Programmed" decisions,* for which guidelines exist (policies, procedures, rules, standards, methods) require less time and analysis to make. However, the development of suitable guidelines for programmed decisions requires thorough analysis.

5. *Decisions dealing with the following strategic elements* of an organization are of great importance and require appropriate care and deliberation: mission, objectives, goals, strategies, organizational structure, managerial and professional roles, managerial style, and procurement and use of organizational resources.

Each manager should develop a set of decision-making guidelines that provide general standards for how much time and analysis to use in evaluating and making different types of decisions.

STEPS IN DECISION MAKING

The decision process begins and ends with the use of information and judgment in removing the uncertainty of future actions. Decisions do not deal with the past; they deal with the present and the future. A decision maker needs to be creative and skillful and to operate from an experience base in using information in arriving at decisions. Successful decision makers usually follow several steps in reaching a decision:

1. Determination of the framework of the problem or opportunity that requires a decision

2. Collection of the appropriate facts and assumptions bearing on the decision
3. Analysis of the facts and assumptions that impact on the problem or opportunity
4. Development of alternatives that are viable in dealing with the decision
5. Analysis of the alternatives based on an assessment of the strengths (benefits) and weaknesses (costs) relative to existing organizational purposes (mission, objectives, goals, and strategies)
6. Selection of the alternative that has the best fit with organizational purposes
7. Delineation of a plan of action on how the decision will be implemented.

These seven steps, in total, formulate the decision-making process carried out by managers and professionals. These steps are found with varying degrees of formality where managers and professionals make decisions in organizations. Decision making requires demanding knowledge, experience, skills, and attitudes in assessing the risks and uncertainties surrounding decisions. Decision situations where the conditions of certainty exist are rare. The decision maker seldom has full knowledge of all the alternatives and their consequences. The search for alternatives continues until the decision maker finds an alternative that best suits some personally determined minimum acceptable level. The notion of trying to find the "best" alternative is controversial in the pragmatic real world.[15] Thus a decision maker searches for alternatives only as long as he is dissatisfied with the best alternative choice he has. There is no real attempt made to put the alternatives in a rank order based on value. A favored alternative is established early in the decision process and survives through to the final formal decision. Early conclusions concerning the selection are often based on subjective or intuitive factors. There is no question that many managers make decisions with an explicit or implicit desire to "get by" and make a decision that is good enough for the situation.

Many of the decisions contain so many complex variables that it is difficult for an individual to examine all of them fully. In real life, managers act within what Herbert Simon has called *bounded rationality*: that is, they make logical decisions, but these decisions are limited by inadequate information and by the ability of the decision maker to utilize that

[15] For instance, see J. M. Roach, "Simon Says . . . Decision Making Is a 'Satisficing' Experience," *Management Review*, **68**(1) (January 1979):8–17.

information. Rather than seeking the best or ideal decision, managers frequently settle for a decision that will adequately serve their purposes. In other words, they "satisfice" or accept the first satisfactory decision they uncover rather than maximizing or searching until they find the optimal decision.[16]

Individuals can develop a facility in making decisions by following several key steps. *First,* they need to develop in their own minds and in their modus operandi a rational approach to the decision-making process. *Second,* having developed a rational approach to the decision-making process, they should proceed methodically and carefully in their collection and analysis of information and in the delineation of alternatives for the decision. *Third,* they should get people involved who can help them in the analysis of the information for the decision and in the assessment of the probability of the decision being effectively implemented.

Delegation of a Decision-Making Authority

Effective managers delegate decision-making authority and responsibility to subordinates. An effective manager does not try to make every decision that is brought to them by subordinates, superiors, and peers. The prudent manager conserves his time and energy for those decisions which fall within his decisionmaking authority and responsibility. Every decision that is brought to a manager should be addressed in the context of: Is this a decision I should make or does somebody else have the responsibility for making this decision within the organization?

It is important that the manager learn how to identify those decisions on problems/opportunities which he must make, those decisions which can be pushed up to the next higher level, and those decisions which have been delegated. Pushing up decisions to the superior should not be simply a passing of the buck, but it should be done in such a way that it represents a clear inability to make the decision at the manager's level, and a clear mandate for the decision to be made by the higher-level person.

Some guidelines that can be used to determine if the decision should be pushed up to the next higher level include answers to such questions as these:

- Are other departments of the organization affected?
- Will that decision have a significant impact on the supervisor's area of responsibility?

[16] Herbert A. Simon, *Models of Man: Social and Rational* (New York: Wiley, 1957). See also James G. March and Herbert A. Simon, *Organizations* (New York: Wiley, 1958).

- Does it require information that is only available at higher levels?
- Does it involve a significant departure from existing strategy?
- Does this problem rest within the delegation of authority and responsibility to this position?

Conflict in Decision Making

Robert Katz states, "Each decision must strike a balance among so many conflicting values, objectives and criteria that it will be suboptimal from any single standpoint. Every decisions or choice affecting the whole enterprise has negative consequences for some of the parts."[17] The fact that so many conflicting values, objectives, and criteria do exist necessitates that the decision maker view the organization to be affected by the decision from a systems perspective, and take into consideration the potential impact of the decision on the stakeholders. The decision maker who occupies a managerial position is always involved in trade-offs in decisions. Two fundamental trade-offs exist: (1) the need to preserve the overall effectiveness in the organization; and (2) the resolution of the inherent conflict found in organizations because people tend to view the organization, and the decision affecting it, from a provincial viewpoint. The manager has to take these provincial viewpoints into consideration and make a decision which best preserves the overall effectiveness of the organization.

Considering the opportunity for conflict and the need for trade-offs, many managers procrastinate and allow these considerations to paralyze their decision making. These are people who would prefer to let things drift with the tide and not make a decision; however, these are the situations in which deliberately not making a decision is in fact a decision itself. It means that resources will continue to be consumed in the way that they had been consumed in the past. Thus, no decision is in fact a decision.

CHOICE ELEMENTS OF DECISION MAKING

Sometimes people make decisions on an informal assessment of alternatives. Such decisions are easy to make. Good decisions are more difficult. Decision making is never totally rational; neither is human behavior—and

---

[17] Robert L. Katz, *Management of the Total Enterprise* (Englewood Cliffs, N.J.: Prentice-Hall, 1970), p. 13.

people make decisions, not systems. Each decision situation faced by a decision maker contains four basic choice elements:

1. What the decision maker *wants* to do
2. What *can be done* considering available resources
3. What *should be done* to satisfy ethical and moral obligations
4. What *must be done* to satisfy existing obligations such as allocating resources for pursuit of existing missions, objectives, goals, and strategies

Table 7-1 contains some examples of each of these choice elements. However rational the decision process becomes, the decision maker will be influenced by these choice elements. Each choice based on an aspect of "can do," "must do," "want to do," and "should do," if taken by itself, could lead the organization in a different direction in its future. If an organizational decision maker chooses a course of action based on personal desires, organizational ineffectiveness could result; for example, if a design engineer chooses a product design that he prefers, but which suboptimizes quality standards, the opportunity for product failure is enhanced. If a manufacturing manager neglects OSHA criteria in the design of production work stations, the opportunity for injury to the workers is increased. Selecting a course of action which does not support existing organizational strategies will create confusion in the organization. If the organization's capabilities cannot support a new strategic direction of the enterprise, then either the capabilities have to be brought into line with the new strategic direction or that direction has to change.

In particular, when analyzing key decisions, the decision maker should be satisfied that each of these four elements is evaluated in the decision process. The decision maker should ask decision-related questions such as the following:

- Have organizational strengths and weaknesses (resources, capabilities, skills, capacities) been considered in his decision? If not, why not?
- Does this decision respect the obligations, commitments, strategies, and stakeholder claims of this organization? If not, why not?
- Have personal preferences been unduly influential in selecting the organization's course of action? If so, will this subjectivity be detrimental to organizational efficacy?
- Have acceptable professional, ethical, and moral standards been respected and maintained in this decision? If not, why not?

Table 7-1. Choice Elements of Organizational Decision Making.

CAN DO	MUST DO	WANT TO DO	SHOULD DO
Organizational facilities: capabilities, capacities, quality, and quantity of personal skills	Satisfy existing contractual obligations	Enhance personal career	Adhere to a code of professional ethics in personal philosophy of management
Quality of leadership	Support of strategies (mission, objectives, goals) in place	Strengthen personal power base in the organization	Select a course of action which enables people to match personal objectives with organizational objectives
Motivation of key people	Legal framework in which competing: fair trade, antitrust laws, etc.	Provide decisive leadership	Contribute to the well-being of society to which you belong
Present competitive position: market share, pricing, costs, quality, etc.	Satisfaction of stakeholder claims	Select the course of action which best matches my value system	Maintain respect for and dignity of people affected by the decision
Financial health	Become and remain profitable in the long term	Select the course of action with the least risk	Be calculative, rational, and objective in decision making

A decision maker cannot separate his decision process from the organizational ambience where the decision is made and where it is executed. People who share resources with the decision maker, if given the opportunity to participate in decision making, can help to evaluate the "can do," "must do," "want to do," and "should do" elements of the decision.

There are limitations in the ability of humans to make purely objective decisions. No decision maker ever has all of the information with which to make a decision. Rather the decisions are made based upon "bounded rationality"; that is, the decision maker collects relevant information and makes the decision within the boundaries established by that information. Another limitation facing the decision maker is simply the inability to gather information directly. Rather the decision information is based upon what has been submitted by many others. The higher up in the organization the decision maker happens to be, the more that decision maker becomes dependent upon underlings for the information.

The decision maker makes the decision based in part upon his own self-interest and to ward off any challenges to his decision-making authority. Information that a manager needs for making decisions, may very well be hidden, might not be relayed, or may be altered in such a way that it is distorted. Political behavior on the part of people will often result in "satisficing" rather than maximize benefits from the decisions. The decision maker always is influenced by his values and experience background.

It is impossible to separate the decision maker from how that person feels the decision will affect his or her own career. Given a choice, a decision maker will make a decision based upon what would most favor his continued position within the organization.

Professor Koontz et al. state that managers have three bases for decision open to them—*experience, experimentation,* and *research and analysis*. They believe that reliance on past experience probably plays a larger part than it deserves in decision making.[18] Experimentation is an obvious way to decide upon alternatives—to try them and see what happens. The experimentation method is costly, particularly when heavy expenditures in capital and personnel are necessary to try a program or a product before a final decision to produce is made. A firm may test a new product in a certain market before expanding its sale nationwide. Research and analysis is a way of testing a decision by searching for relationships among the critical variables, constraints, and premises that bear upon the objective being sought. The solution of the planning problem using research and analysis requires that it be broken down into its component

[18] Harold Koontz, C. O'Donnel, and H. Weihrich, *Management* (New York: McGraw-Hill, 1980), p. 240.

parts and the various tangible and intangible factors studied. A major characteristic of the research and analysis technique is to use models simulating the problem, issue, or opportunity. It is less expensive to make mistakes on paper.

In part, decision making involves a process of "muddling through" in addition to "satisficing." Muddling through is the process of making piecemeal or incremental decisions one after the other as the course of action develops and its anticipated and unanticipated consequences are evaluated.[19]

DECISION MAKING WITHIN MANAGEMENT FUNCTIONS

Each of the management functions inherently contains many questions which require decisions. Making a decision and answering these questions are vital to the effective discharge of these functions. Table 7-2 contains just a sample of the many questions that could be answered about a particular management function.

Planning entails the most significant and far-reaching decisions the decision maker can make. In the planning process, the decision maker establishes the mission, objectives, goals, and strategies for the organization, delineating how resources will be used and how people will be assigned in the use of those resources.

There is an old management saw that says: A well-defined problem is already well on toward its solution. Organizational problems, issues, and opportunities are seldom simple and isolated. All parts of an organizational system interact; all facets of an organizational decision situation interact as well. A decision which results in a poor product design impacts the product itself and

- Product quality assurance
- Product reliability
- Product maintainability
- Product costs (it costs to produce the poorly designed product *and* then to correct the design deficiency)
- Product service
- Technical documentation
- Corporate image
- Spare parts
- Product producing, marketing, finances, and future product design and development

[19] Herbert A. Simon, *Administrative Behavior*, 3rd ed. (New York: Macmillan, 1976), pp. 38–41, 80–83, 240–244.

Table 7-2. Decisions of the Team Management Functions.[a]

TEAM PLANNING

1. What is the mission or "business" of the team?
2. What are the team's principal objectives?
3. What team goals must be attained in order to reach team objectives?
4. What is the strategy that will be used by the team to accomplish its purposes?
5. What resources are available for the team's use in accomplishing its mission?

TEAM ORGANIZATION

1. What is the basic organizational design of the team?
2. What are the individual and collective roles on the team that must be identified, defined, and negotiated?
3. Will the team members understand and accept the authority, responsibility, and accountability that is assigned to them as individuals and as a team?
4. Do the team members understand their authority and responsibility to make decisions?
5. How can the team effort be coordinated so that the members will work in harmony, not against one another?

TEAM MOTIVATION

1. What motivates the team members to do their best work?
2. Does the team manager provide a leadership style acceptable to the members of the team?
3. Is the team "productive"? If not, why not?
4. What can be done to increase the satisfaction and productivity of the team members?
5. Are the team meetings conducted in such a manner that people attending are encouraged? Discouraged?

TEAM DIRECTION

1. Is the team leader qualified to lead the team?
2. Is the team leader's style acceptable to the members of the team?
3. Do the individual members of the team assume leadership in the areas where they are expected to lead?
4. Is there anything that the team leader can do to increase the satisfaction of the team members?
5. Does the team leader inspire confidence, trust, loyalty, and commitment among the team members?

TEAM CONTROL

1. Have performance standards been established for the team? For the individual members?
2. What feedback on the team's performance does the manager who appointed the team receive?
3. How often does the team get together to formally review its progress?
4. Has the team attained its objectives and goals in an effective and efficient manner?
5. Do the team members understand the nature of control in the operation of the team?

[a] Adapted from M. H. Mescon et al., *Management* (New York: Harper & Row, 1981), p. 167.

In large organizations there may be hundreds of interrelated facets affected by a key decision. The prudent decision maker, in the rational approach to decision making, "thinks through" the "systems effects" of a course of action during the decision process. Decisions made in a team ambience helps to identify the potential systems effects of the decision.

Team or group decisions conducted in a laboratory setting are often more accurate than individual members working alone.[20] Team decision making usually takes more time than an individual working alone. However, people are more willing to support a decision that they have helped to make. Under what conditions would a decision maker seek participation in making decisions

- When the identification, collection, and analysis of information can best be done by a team of people?
- When the problem, issue, or opportunity is of an unstructured, systems nature?
- When the stakeholders' acceptance is crucial to implementation of the decision?
- When the knowledge and skills of the team members are crucial to effective analysis of the decision?
- When it is the policy of the organization to seek genuine participation, through a consensus process, of people in those decisions which affect them?
- When there is a high degree of compatibility between the goals of the participants and the organization?[21]

Sims and Dean see self-managing teams as logical extensions of quality circles. Such teams have as their theme placing a high degree of decision-making responsibility and behavior control in the hands of the team itself. The team is the work group usually undertaking the following responsibilities beyond the traditional "bottom-rung" jobs:

> (1) preparing an annual budget, (2) taking on timekeeping responsibilities, (3) recording quality control statistics, (4) compiling in-process inventory statistics, (5) making within-group job assignments, (6) solving technical problems, (7) training fellow team members, (8) adjusting production schedules, (9) redesigning and modifying production pro-

[20] James H. Davis, *Group Performance* (Reading, Mass.: Addison-Wesley, 1969); Jay Hall, "Decisions, Decisions, Decisions, Decisions," *Psychology Today*, November 1971, p. 51.
[21] Paraphrased in part from Victor H. Vroom, "A New Look at Managerial Decision Making," *Organizational Dynamics*, Spring 1973, pp. 1–15.

cesses, (10) setting team goals, (11) resolving internal conflicts, and (12) assessing internal performance.[22]

Success in such teams require such obvious factors as top-management support, adequate role definition, and defined task boundaries. Feedback is also required so the teams know how well they are able to accomplish the things they set out to do.

In managing teams the need for participation in decision making is clear. Team leaders and team members should be aware of some of the difficulties of team decision making:

1. Inappropriate membership on the team.
2. Lack of relevant information for team analysis.
3. Lack of interest or weak concern by team members.
4. Perceptions of team members that their participation is meaningless and that the decision maker will act without really listening to their recommendations.
5. Danger of interpersonal strife developing during the discussions about the decision. If the team interactions confront people rather than problems, an unsavory cultural ambience can develop which will adversely affect the quality of the decisions.
6. Dominance by the team leader or other members of the team which inhibits the participation of less assertive individuals.
7. Inadequate team leadership which fails to define issues and decision parameters, encourage analysis, summarize progress (or lack of progress) or focus team efforts and bring the decision to a closure point.
8. The dangers of *Groupthink,* a ". . . deterioration of mental efficiency, reality testing, and moral judgment that results from in-group pressures."[23] Some of the symptoms of *Groupthink* include:

    - Superior feelings regarding outsiders
    - Willingness to take extreme risks based on feelings of superiority, and rationalizations that discount warning of risks
    - Unrealistic and stereotyped thinking about outsiders and outside conditions
    - Suppression of dissent within the group
    - Belief that dissent is disloyal

[22] Henry P. Sims, Jr., and James W. Dean, Jr., "Beyond Quality Circles: Self-Managing Teams," *Personnel,* January 1985, p. 26.
[23] See Irving L. Janis, *Victims of Groupthink* (Boston: Houghton Mifflin, 1972), p. 9.

- A belief concerning issues that silence means consent
- Self-appointed "mindguards" who protect the group from adverse information

Groupthink damages the rational decision process; the zeal for consensus may go beyond its enthusiasm for effective decisions. *Groupthink* occurs to some degree in all groups; it is more likely to occur in highly cohesive groups which are isolated from outside and conflicting opinions, and under leaders who promote a preferred solution.[24]

How can team decision making be improved? If the team is composed of the right persons, and team mission, objectives, goals, and strategies are in place, then a qualified chairperson has a better chance of success on consensus decision making. A competent chairperson can help in facilitating a favorable ambience for decision making through doing the following:

1. Acquaint the team members with Douglas McGregor's description of effective and ineffective groups contained in the below quote. If the team sees the relationship between McGregor's effective groups and prudent team decision making, an important step has been taken to facilitate effective decision making.[25]

   "Let us attempt to lay aside for a moment our prejudice, pro or con, with respect to group activity and consider in everyday common-sense terms some of the things that are characteristic of a well-functioning, efficient group. Occasionally, one does encounter a really good top management team or series of staff meetings or committee. What distinguishes such groups from other less effective ones?

   a. The "atmosphere," which can be sensed in a few minutes of observation, tends to be informal, comfortable, relaxed. There are no obvious tensions. It is a working atmosphere in which people are involved and interested. There are no signs of boredom.
   b. There is a lot of discussion in which virtually everyone participates, but it remains pertinent to the task of the group. If the discussion gets off the subject, someone will bring it back in short order.

---

[24] Ibid.

[25] Douglas McGregor, *The Human Side of Enterprise* (New York: McGraw-Hill, 1960), pp. 232–238.

c. The task or the objective of the group is well understood and accepted by the members. There will have been free discussion of the objective at some point until it was formulated in such a way that the members of the group could commit themselves to it.
d. The members listen to each other! The discussion does not have the quality of jumping from one idea to another unrelated one. Every idea is given a hearing. People do not appear to be afraid of being foolish by putting forth a creative thought even if it seems fairly extreme.
e. There is disagreement. The group is comfortable with this and shows no signs of having to avoid conflict or to keep everything on a plane of sweetness and light. Disagreements are not suppressed or overridden by premature group action. The reasons are carefully examined, and the group seeks to resolve them rather than to dominate the dissenter.

   On the other hand, there is no "tyranny of the minority." Individuals who disagree do not appear to be trying to dominate the group or to express hostility. Their disagreement is an expression of a genuine difference of opinion, and they expect a hearing in order that a solution may be found.

   Sometimes there are basic disagreements which cannot be resolved. The group finds it possible to live with them, accepting them but not permitting them to block its efforts. Under some conditions, action will be deferred to permit further study of an issue between the members. On other occasions, where the disagreement cannot be resolved and action is necessary, it will be taken but with open caution and recognition that the action may be subject to later reconsideration.
f. Most decisions are reached by a kind of consensus in which it is clear that everybody is in general agreement and willing to go along. However, there is little tendency for individuals who oppose the action to keep their opposition private and thus let an apparent consensus mask real disagreement. Formal voting is at a minimum; the group does not accept a simple majority as a proper basis for action.
g. Criticism is frequent, frank, and relatively comfortable. There is little evidence of personal attack, either openly or in a hidden fashion. The criticism has a constructive flavor in that it is oriented toward removing an obstacle that faces the group and prevents it from getting the job done.
h. People are free in expressing their feelings as well as their ideas

both on the problem and on the group's operation. There is little pussyfooting, there are few "hidden agendas." Everybody appears to know quite well how everybody else feels about any matter under discussion.

i. When action is taken, clear assignments are made and accepted.

j. The chairman of the group does not dominate it, nor on the contrary, does the group defer unduly to him. In fact, as one observes the activity, it is clear that the leadership shifts from time to time, depending on the circumstances. Different members, because of their knowledge or experience, are in a position at various times to act as "resources" for the group. The members utilize them in this fashion and they occupy leadership roles while they are thus being used.

   There is little evidence of a struggle for power as the group operates. The issue is not who controls but how to get the job done.

k. The group is self-conscious about its own operations. Frequently, it will stop to examine how well it is doing or what may be interfering with its operation. The problem may be a matter of procedure, or it may be an individual whose behavior is interfering with the accomplishment of the group's objectives. Whatever it is, it gets open discussion until a solution is found.

Now let us look at the other end of the range. Consider a poor group—one that is relatively ineffective in accomplishing its purposes. What are some of the observable characteristics of its operation?

a. The "atmosphere" is likely to reflect either indifference and boredom (people whispering to each other or carrying on side conversations, individuals who are obviously not involved, etc.) or tension (undercurrents of hostility and antagonism, stiffness and undue formality, etc.). The group is clearly not challenged by its task or genuinely involved in it.

b. A few people tend to dominate the discussion. Often their contributions are way off the point. Little is done by anyone to keep the group clearly on the track.

c. From the things which are said, it is difficult to understand what the group task is or what its objectives are. These may have been stated by the chairman initially, but there is no evidence that the group either understands or accepts a common objective. On the contrary, it is usually evident that different people have different, private, and personal objectives which they are attempting to achieve in the group, and that these are often in conflict with each other and with the group's work.

d. People do not really listen to each other. Ideas are ignored and overridden. The discussion jumps around with little coherence and no sense of movement along a track. One gets the impression that there is much talking for effect—people make speeches which are obviously intended to impress someone else rather than being relevant to the task at hand.

   Conversation with members after the meeting will reveal that they have failed to express ideas or feelings which they may have had for fear they would be criticized or regarded as silly. Some members feel that the leader or the other members are constantly making judgments of them in terms of evaluations of the contributions they make, and so they are extremely careful about what they say.

e. Disagreements are generally not dealt with effectively by the group. They may be completely suppressed by a leader who fears conflict. On the other hand, they may result in open warfare, the consequence of which is domination by one subgroup over another. They may be "resolved" by a vote in which a very small majority wins the day, and a large minority remains completely unconvinced.

   There may be a "tyranny of the minority" in which an individual or a small subgroup is so aggressive that the majority accedes to their wishes in order to preserve the peace or to get on with the task. In general only the more aggressive members get their ideas considered because the less aggressive people tend either to keep quiet altogether or to give up after short, ineffectual attempts to be heard.

f. Actions are often taken prematurely before the real issues are either examined or resolved. There will be much grousing after the meeting by people who disliked the decision but failed to speak up about it in the meeting itself. A simple majority is considered sufficient for action, and the minority is expected to go along. Most of the time, however, the minority remains resentful and uncommitted to the decision.

g. Action decisions tend to be unclear—no one really knows who is going to do what. Even when assignments of responsibility are made, there is often considerable doubt as to whether they will be carried out.

h. The leadership remains clearly with the committee chairman. He may be weak or strong, but he sits always "at the head of the table."
i. Criticism may be present, but it is embarrassing and tension-producing. It often appears to involve personal hostility, and the members are uncomfortable with this and unable to cope with it. Criticism of ideas tends to be destructive. Sometimes every idea proposed will be "clobbered" by someone else. Then, no one is willing to stick his neck out.
j. Personal feelings are hidden rather than being out in the open. The general attitude of the group is that these are inappropriate for discussion and would be too explosive if brought out on the table.
k. The group tends to avoid any discussion of its own "maintenance." There is often much discussion after the meeting of what was wrong and why, but these matters are seldom brought up and considered within the meeting itself where they might be resolved.

2. Undertake the role of a "facilitator" whose purpose is to help the group exchange ideas and viewpoints, nurture the group process, getting salient issues out for discussion and encouraging everyone's participation by drawing out the less vocal members.
3. Keep in mind one of the key roles of any manager/leader: facilitating an environment where people work together to attain common purposes and gain social, economic, and self-satisfaction in so doing.
4. Recognize that a team decision process is quite similar to the steps in decision process described on pages 237–238. By agreeing to and following these steps the team members can see how the process they are following fits together. If the team wishes, this process can be modified to suit their own needs. The important thing is that the team follow a process.
5. Encourage participation by all team members to debate, test hypotheses, act in adversial roles to test opinions and results, and participate in "brainstorming" to elicit a free flow of ideas.
6. Suggest reexamination of a decision if it appears that the solution was reached prematurely, or through a *Groupthink* dominance. If individual members sense an insufficient generation of alternatives, or insufficient critical assessment of certain alternatives, then encourage the team to break away from the problem to do further study and analysis. Then reconvene the team and proceed with the decision process. Sometimes new ideas and approaches need an

incubation period; an alert chairperson will sense this and allow time for the necessary incubation.

7. Take a little time at the beginning of the team decision deliberations to discuss how the different disciplines represented by the team members can, if not considered, create disciplinary conflicts that may impair the efficiency of the decision process. If the team members recognize this risk and the need for better communication, the disciplinary conflicts can be minimized.

MANAGEMENT SCIENCE

Management science adds the application of scientific methodology to the study of alternatives in a problem situation with a view to use quantitative bases for arriving at a solution. In terms of the objectives or goals sought, this is the hallmark of the management science or operations research school. Operations research, however, just like financial and accounting analysis, does not make the decision, but rather provides a quantitative data base that can help the manager make the decision. The operations researchers emphasize defining the problem and goals, carefully collecting and evaluating data, developing and testing hypotheses, determining relationships among data, developing and checking predictions based upon hypotheses, and devising measures to evaluate the effectiveness of a course of action. The characteristics of the operations research approach as applied to decision making may be summarized as follows:

1. The emphasis on models—the logical representation of a reality or problem. These may, of course, be simple or complex. For example, the accounting formula "assets minus liabilities equals proprietorship" ($A - L = P$) is a model, since it represents an idea and, within the limits of the terms used, symbolizes the relationship among the variables involved.
2. The emphasis on goals (in a problem area and the development of measures of effectiveness in determining whether a given solution shows promise of attaining the goal. For example, if the goal is profit, the measure of effectiveness may be the rate of return on investment, and every proposed solution will arrange the variables so that the end result can be weighed against this measure. Some variables may be subject to control by the manager; others may represent uncontrollable factors in the system.
3. The attempt to incorporate in a model the variables in a problem, or at least those which appear to be important to its solution.

4. Putting the model with its variables, constraints, and goals in mathematical terms so that it may be clearly perceived, subjected to mathematical simplification, and readily utilized for calculation by substitution of quantities for symbols.
5. The attempt to quantify the variables in a problem to the extent possible, since only quantifiable data can be inserted into a model to yield a finite result.
6. The attempt to supplement quantifiable data with such usable mathematical and statistical devices as the probabilities in a situation, thus making the mathematical and computing problem under uncertainty workable within a small and perhaps insignificant margin of error.[26]

The evaluation of alternatives is an important step in the decision-making process. Quantitative techniques provide a basis for better assessment of alternatives through providing for the development and use of models to evaluate risk, uncertainty, and complexity in decision problems. Quantitative analysis provides a rational approach in the systematic application of scientific methods to the solution of management decisions. Quantitative techniques offer an alternative to intuitive "seat-of-the-pants" decision making; these techniques provide the opportunity for the decision maker to sharpen his judgment in making decisions. The final decision is always based on judgment.

Management science is often viewed solely as a mathematical or quantitative approach to decision making. It is much more: a totality of many philosophies and techniques used in a systematic way to help the decision maker make decisions. Management science is characterized by the process shown in Figure 7-1.

The steps shown in Figure 7-1 are not always followed in the orderly fashion portrayed in the figure. The initial definition may be changed several times before the issues are accurately identified. Other steps in the approach may be changed as feedback improves the decision maker's understanding of the problem. A brief summary of each of the steps follows.[27]

Orientation: At this stage the problem is identified and its characteristics and boundaries are defined in broad terms.

Problem definition: In this stage the decision problem is expressed in both qualitative and quantitative terms.

[26] Koontz et al., op. cit., p. 250.

[27] Paraphrased from D. Cleland and D. F. Kocaoglu, *Engineering Management* (New York: McGraw-Hill, 1981), pp. 184–200.

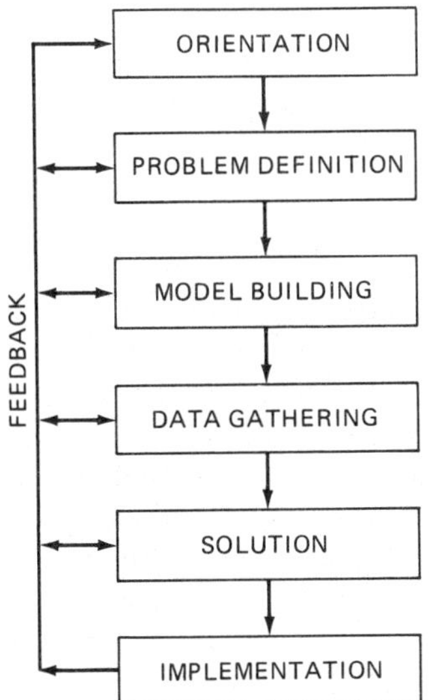

Figure 7-1. Management science approach. (*Source:* David I. Cleland and Dundar F. Kocaoglu, *Engineering Management* (New York: McGraw-Hill, 1981) p. 184.)

Model building: Model building is the development of a mathematical description of the decision problem to reflect the characteristics of the issues involved.

Data gathering: The data are gathered after the model is developed and checked for internal consistency. Generally there is an abundance of data in organizations. However, the existence of data does not necessarily mean that the quantitative information is readily available to solve the model. It is necessary to identify the useful information, to separate the relevant and irrelevant data, and to concentrate on obtaining the data required by the model.

Solution: The solution of mathematical models can be obtained in several ways. Graphical solution is the simplest approach for problems with low dimensionality, but it is not a viable method for models with more than two or three variables. Many decision problems require the solution of simultaneous equations. They can be solved manually or by computer, using one of the several reduction techniques available. Calculus is used for the solution of simple optimization models, but special optimization techniques are necessary to solve complex problems. Most optimization methods utilize iterative procedures or systematic search techniques. Computers play an important role in their solution.

Implementation: Solving the model is not the end of the management science approach. Unless the results are put to use, the effort spent in developing the model and finding the solution will have been wasted. The most meaningful test of a model lies in its acceptability and implementation. Even highly sophisticated and accurate models would be useless if they were not implemented.

The management science school uses decision theory in the process of making decisions, claiming that it is more of a science than an art. Management science decision theory models focus on the elements in decision situations which are common to all decisions, and which can help the decision maker use a framework which enables the analysis of a complex situation with numerous alternatives and possible consequences.

In the material that follows, we will acquaint the reader with management science approaches; we will not concern ourselves with the methodology for each technique. A decision maker who wishes to use these techniques can find the methodology outlined in some of the references in the end-of-chapter bibliography.

Linear Programming uses graphic, algebraic, or simplex methodologies to design a schema to optimize the allocation of scarce resources. The technique is built around two or more activities competing for limited resources; a linear relationship is assumed in the problem and the objective. The technique is useful when input data is quantifiable and objectives are measurable. Logistics or operating facilities issues such as warehousing and transportation are solved by linear programming.

Queuing theory, or waiting-line theory, deals with the issue of balancing the cost of having a waiting line against the cost of service to maintain that line in servicing a system. Queuing theory seeks to minimize employee idle time while providing adequate client service. Use of airline reservation clerks, data waiting for computer processing, trucks waiting to unload cargo, and airplanes waiting to use docking facilities are examples of where queuing theory methodology can be useful.

Probability theory uses statistics to help the decision maker reduce risk. By predicting past patterns, a decision maker can improve current and future decisions. Using statistics to evaluate the probability of an event occurring sharpens a decision maker's judgment. Probability ranges from 1.0, when an event is likely to occur, to 0 when the event is certain not to occur. Probability can be determined objectively, or based on historical trends, or by subjective assessment by the decision maker based on past experience.

Marginal analysis deals with the additional cost in a particular decision, rather than the average cost. Marginal analysis helps to make decisions in optimizing returns or minimizing costs. Decisions concerning taking on a new customer could be studied using marginal analysis to determine what additional revenue would be generated by that new order. Marginal analysis helps to answer the question, Do the incremental benefits exceed the incremental costs of an action?

Inventory models are used to minimize the costs of ordering and carrying inventory, against the danger of being out of stock. By estimating overall demand for a particular period, order placement costs, carrying costs, and item value, one can determine an optimum size for efficient order quantities. The notion of inventory models is being modified by Japanese "just-in-time" (JIT) production/inventory management. JIT means to produce and deliver finished goods just in time to be sold, subassemblies just in time to be assembled into finished goods, and purchased materials just in time to be transformed into fabricated parts.[28]

Breakeven models are concerned with the relationship between total costs and total revenues. When revenues cover variable costs and the increment left over is equal to fixed costs, the breakeven point has been reached.

Network analysis (CPM, PERT) provides methodology for scheduling activities to show interrelationships and meet target goals. Charts developed through network analysis help the decision maker to make planning and control decisions and to highlight where resource reallocation needs to be carried out.

Simulation uses models to imitate an operation or situation when the variables, components, and relationships of a system are known. Components of the decision problem are mathematically defined and related to each other in a series of functional relationships. A proposed change, such as services, personnel, or location, can be "simulated" and the result measured. Simulation models can be used for problems that are too complex to be described or solved by standard mathematical equations. These models replicate an operation to see what will happen over time as variables are changed. Simulation models are also used to test machine systems and large-scale operations such as airports or other transportation systems.

Return on investment is a popular financial criterion to measure the productivity of assets. Profit as a percentage of capital investment in an organization is measured; it is then possible to determine how well the investment is being used to generate profits. Other financial methods for measuring the productivity of resources are described in any good financial/accounting book.

---

[28] See Richard J. Schonberger, *Japanese Manufacturing Techniques* (New York: Free Press, 1982).

TRADE-OFF ANALYSIS

Successful team management is both an art and a science and attempts to control corporate resources within the constraints of time, cost, and performance. Most engineering projects are unique, one-of-a-kind activities for which there may not have been reasonable standards for forward planning. As a result, the team manager may find it extremely difficult to stay within the time-cost-performance triangle of Figure 7-2.

The time, cost, and performance triangle is that "magic combination" which is continuously pursued by the project manager throughout the life cycle of the project. If the project were to flow smoothly, according to plan, there may not be any need for trade-off analysis. Unfortunately, most projects eventually find crises where this delicate balance necessary to attain the desired performance within time and cost is no longer possible.

This is shown in Figure 7-3, where the Greek deltas represent the deviations from the original estimates. The time and cost deviations are normally overruns, whereas the performance error will be an underrun. No two projects are ever exactly alike, and trade-off analysis would appear to be an ongoing effort throughout the life of the project, continuously influenced by both the internal and external environment. Experienced team managers may have predetermined trade-offs in reserve as appropriate crises arise. Experienced team managers recognize that trade-offs are a continuous thought process.

Trade-offs are always based upon the constraints of the project. Table 7-3 illustrates the types of constraints commonly imposed. Situations A and B are the typical trade-offs encountered in project management. For example, situation A-3 portrays most research and development projects.

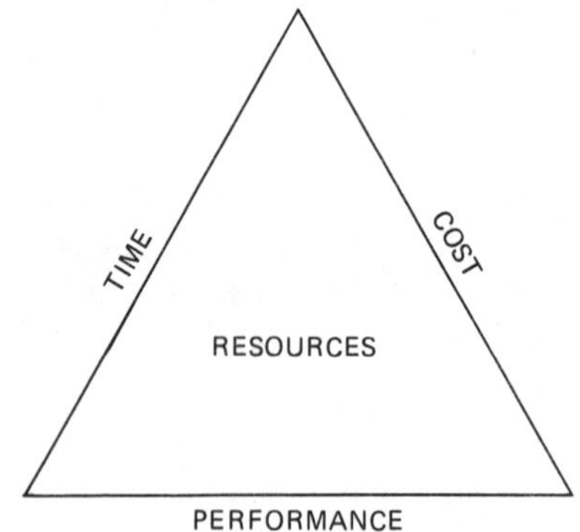

Figure 7-2. Overview of project management.

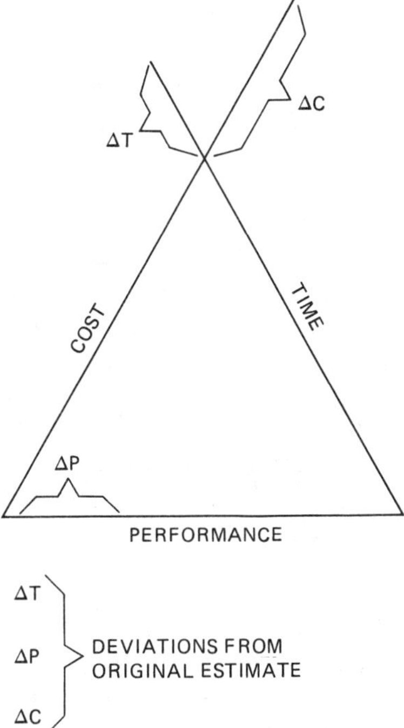

Figure 7-3. Project management with trade-offs.

Table 7-3. Categories of Constraints.

	TIME	COST	PERFORMANCE
A. ONE ELEMENT FIXED AT A TIME			
A-1	Fixed	Variable	Variable
A-2	Variable	Fixed	Variable
A-3	Variable	Variable	Fixed
B. TWO ELEMENTS FIXED AT A TIME			
B-1	Fixed	Fixed	Variable
B-2	Fixed	Variable	Fixed
B-3	Variable	Fixed	Fixed
C. THREE ELEMENTS FIXED ON VARIABLE			
C-1	Fixed	Fixed	Fixed
C-2	Variable	Variable	Variable

The performance of an R&D project is usually well defined, and it is cost and time that may be allowed to go beyond budget and schedule. The decision to determine what to sacrifice lies in the available alternatives. If there are no alternatives to the product being developed and the potential usage is great, then cost and time are the tradeoffs.

Most capital equipment projects would fall into situation A-1 or B-2, where time is of the essence. The sooner the piece of equipment gets into production, the quicker the return of investment can be realized. Often there exist performance constraints that determine the profit potential of the project. If the project potential is determined to be great after all alternatives have been established, cost will be the slippage factor as in B-2.

Nonprocess-type equipment, such as air pollution control equipment, usually develops a scenario around situation B-3. Performance is fixed by the Environmental Protection Agency. Although the deadline for compliance can be delayed through litigation, if the lawsuits fail, most firms then try to comply with the least expensive equipment that will meet the minimum requirements.

The professional consulting firm operates primarily under situation B-1. In situation C, the trade-off analysis will be completed based on the selection criteria and constraints. In C-1, if everything is fixed, there is no room for any outcome other than total success, and if everything is variable (C-2), there are no constraints and thus no trade-off.

Many factors go into the decision to sacrifice either time, cost, or performance. It should be noted, however, that it is not always possible to sacrifice one of these items without affecting the others. For example, reducing the time could have serious impact on performance and cost (especially if overtime is required).

There are several factors which tend to "force" trade-offs. Poorly written documents such as statements of work, contracts, and specifications are almost always inward-forcing conflicts where the project manager tends to look for performance relief. In many projects, the initial sale and negotiation, as well as specification writing, are done by highly technical people who are driven to create a monument rather than meet the operational needs of the customer, the operator of the system. When the operating forces dominate outward from the project to the customer, project managers may tend toward cost relief.

METHODOLOGY FOR TRADE-OFF ANALYSIS

Any process for managing time, cost, and performance trade-offs should emphasize the systems approach to management by recognizing that even the smallest change in a project or system could easily affect all of the organization's systems. A typical systems model is shown in Figure 7-4. Because of this, it is often better to develop a process for decision-making/trade-off analysis rather than to maintain hard and fast rules on trade-offs. The following six steps might be a representative method for managing project time, cost, and performance trade-offs:

1. Recognizing and understanding the basis for project conflicts
2. Reviewing the project objectives
3. Analyzing the project environment and status
4. Identifying the alternative courses of action
5. Analyzing and selecting the best alternative
6. Revising the project plan

The foregoing quantitative/management science techniques offer advantages in evaluating alternatives in the decision process. The analysis gained from the use of such techniques is only as good as the data used in the analysis and the assumptions underlying such analysis. A decision maker should understand both the advantages and limitations of such techniques to sharpen the judgment factor in decisions. Yet with their

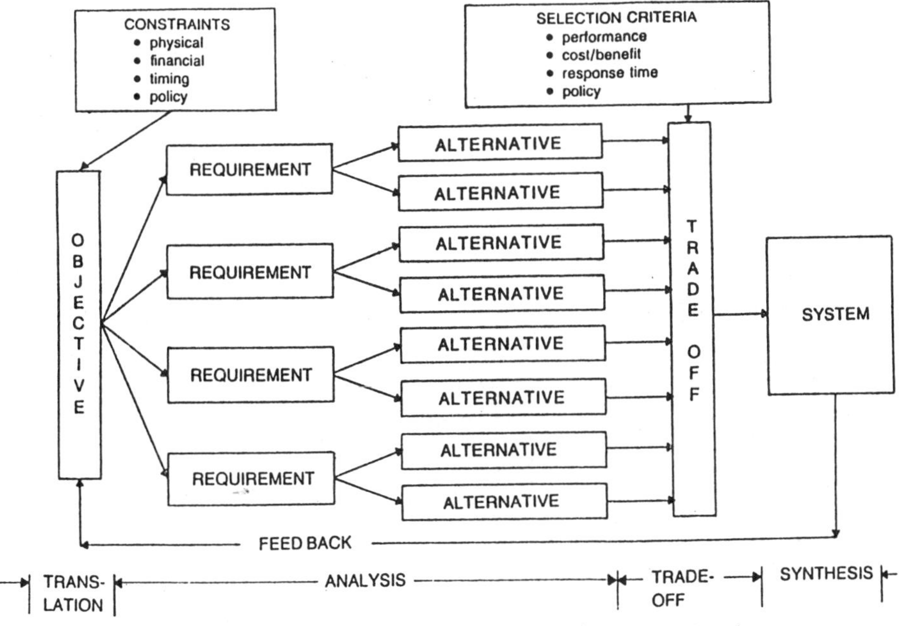

Figure 7-4. The systems approach (from project management).

limitations, such techniques should not be overlooked to facilitate the decision-making process.

SUMMARY

Decision making pervades everything that a manager does. The ability to make decisions is a skill that requires practice, restudy, and more practice. Decision making is closely related to—indeed inescapable from—the management process.

In this chapter we defined a decision and the decision process, noting that problems and opportunities in organizations provide the basis for the making and execution of decisions. Types of decisions were discussed along with a description of the steps in decision making. The delegation of authority and responsibility to make decisions constitutes the basis for decentralization in organizations.

A decision maker faces four basic choice elements: What the decision maker *wants to do; what can be done; what should be done;* and *what must be done*. These choice elements create conflict for the decision maker who must draw on experience, experimentation, and research and analysis in selecting a course of action from among alternatives.

We related decision making to the management functions, and by drawing on the work of Douglas McGregor, showed how the ambience of a team's meeting can influence the effectiveness of the group. We concluded the chapter by briefly reviewing the contributions of management science to decision-making philosophies and methodologies.

In Chapter 8 we present material on time management, stress, and burnout.

BIBLIOGRAPHY

Agor, Weston H. "Using Intuition to Manage Organizations in the Future." *Business Horizons*, July–August 1984.

Cleland, David I., and Dundar F. Kocaoglu, *Engineering Management*, New York: McGraw-Hill, 1981.

Davis, James H. *Group Performance*. Reading, Mass.: Addison-Wesley, 1969.

Drucker, Peter F. *The Practice of Management*. New York: Harper & Row, 1954.

Drucker, Peter F. *Managing for Results*. New York: Harper & Row, 1964.

Hall, Jay. "Decisions, Decisions, Decisions, Decisions." *Psychology Today*, November 1971.

Hampton, David R. *Contemporary Management*. New York: McGraw-Hill, 1981.

Janis, Irving, L. *Victims of Groupthink*. Boston: Houghton Mifflin, 1972.

Katz, Robert L. *Management of the Total Enterprise*. Englewood Cliffs, N.J.: Prentice-Hall, 1970.

Knight, F. H. *Risk, Uncertainty, and Profit*. New York: Harper & Row, 1920.

Koontz, Harold, C. O'Donnell, and H. Weihrich. *Management.* New York: McGraw-Hill, 1980.
March, James G, and Herbert A. Simon. *Organizations.* New York: Wiley, 1958.
McGregor, Douglas, *The Human Side of Enterprise.* New York: McGraw-Hill, 1960.
Mescon, M. H. et al. *Management.* New York: Harper & Row, 1981.
Mintzberg, Henry. *The Nature of Managerial Work.* New York: Harper & Row, 1973.
Roach, J. J. "Simon Says . . . Decision Making Is a 'Satisficing' Experience." *Management Review,* vol 68, no. 1, January 1979.
Robbins, Stephen P. *The Administrative Process: Integrating Theory and Practice.* Englewood Cliffs, N.J.: Prentice-Hall, 1976.
Schoderbek, Peter P. *Management Systems,* 2nd ed. New York: Wiley, 1971.
Schonberger, Richard J. *Japanese Manufacturing Techniques.* New York: Free Press, 1982.
Simon, Herbert A. *Models of Man: Social and Rational.* New York: Wiley, 1957.
Simon, Herbert A. *Administrative Behavior,* 3rd ed. New York: Macmillan, 1976.
Simon, Herbert A. *The New Science of Management Decision.* rev. ed. Englewood Cliffs, N.J.: Prentice-Hall, 1977.
Sims, Henry P. Jr., and James W. Dean, Jr. "Beyond Quality Circles: Self-Managing Teams." *Personnel,* January 1985.
Stoner, James A. F. *Management,* 2nd ed. Englewood Cliffs, N.J.: Prentice-Hall, 1982.
Vroom, Victor H. "A New Look at Managerial Decision Making." *Organizational Dynamics,* Spring 1973.

8
TIME MANAGEMENT, STRESS, AND BURNOUT

UNDERSTANDING TIME MANAGEMENT

For most people, time is a resource which, when lost or misplaced, is gone forever. For a project manager, however, time is more of a constraint, and effective time management principles must be employed to make it a resource.

Most executives prefer to understaff projects with the mistaken belief (or should we say, hope) that the project manager will assume the additional work load. Unfortunately, this is easier said than done. The project manager may already be heavily burdened with meetings, report preparation, internal and external communications, conflict resolution, and planning/replanning for crises. And yet each project manager somehow managers to manipulate his time so that the work will get done.

Inexperienced project managers often work large amounts of overtime with the faulty notion that this is the only way to get the job done. While this may be true, experienced personnel soon learn to delegate and how to employ effective time management principles.

The major problem with time management is getting people to realize that there exists a time management problem and that solutions are possible. The following questions should make the reader realize that each of us has room for improvement.

- Do you have trouble completing work within the allocated deadlines?
- How long can you work at your desk before being interrupted? How many interruptions each day?
- Do you have a procedure for handling interruptions?
- If you need a large block of uninterrupted time, is it available? With or without overtime?
- How do you handle drop-in visitors and phone calls?
- How is incoming mail handled?

- Are you accomplishing more or less than you were three months ago? Six months ago?
- How difficult is it for you to say no?
- How do you approach detail work?
- Do you perform work that should be handled by your subordinates?
- Do you have sufficient time each day for personal interests?
- Do you still think about your job when away from the office?
- Do you make a list of things to do? If yes, is the list prioritized?
- Does your schedule have some degree of flexibility?
- Do you have established procedures for routine work?

While it may not be possible to cope with all of these questions, the more one can deal with, the greater the opportunity for the project manager to convert time from being a constraint to becoming a resource.

TIME ROBBERS

Project managers are not merely managers, but are doers as well. As a result, they suffer from the time robbers of both the managers and the doers. If the project manager is not careful and overemphasizes his role of a doer rather than a manager, the impact of the time robbers may become monumental.

The most challenging problem facing the project manager is his inability to say no. Consider a situation in which an employee comes into your office with a problem. The employee may be sincere when he says that he simply wants your advice but, more often than not, the employee wants to take the monkey off of his back and put it on to yours. If the latter is, in fact, the real truth, then the employee's problem is now *your* problem.

The correct way to handle a situation as this is to first screen the problems with which you wish to get involved. Second, if the situation does necessitate your involvement, then you must make sure that when the employee leaves your office, the employee realizes that the problem *is still his,* not yours. Third, if you find that the problem will require your continued attention, then remind the employee that all future decisions will be joint decisions and that the problem will still be on the employee's shoulders. Once employees realize that they cannot put their problems on your shoulders, then they soon learn how to make decisions and *your* time demands may ease up.

There are numerous time robbers in the project management environment.
These include

- Incomplete work
- A job poorly done that must be done over
- Delayed decisions
- Poor communications channels
- Uncontrolled telephone calls
- Casual visitors
- Waiting for people
- Failure to delegate or unwise delegation
- Poor retrieval system
- Lack of information in a ready-to-use format
- Day-to-day administration
- Spending more time than anticipated in answering questions
- Lack of sufficient clerical support
- Late appointments
- Impromptu tasks
- Union grievances
- Having to explain "thinking" to superiors
- Too many levels of review
- Too many people in a small area
- Office casual conversations
- Misplaced information
- Sorting mail
- Record keeping
- Shifting priorities
- Indecision or delaying decisions
- Procrastination
- Proofreading correspondence
- Setting up appointments
- Too many meetings
- Monitoring delegated work
- Unclear roles/jobs descriptions
- Unnecessary crisis intervention
- Overcommitted outside activities
- Executive meddling
- Budget adherence requirements
- Poorly educated customer
- Lack of adequate responsibility and commensurate authority
- Poor functional performance
- Changes without direct notification/explanation
- Lack of authorization to make judgment decisions
- Poor functional status reporting
- Inability to use one's full potential
- Overeducated for daily tasks
- Work overload
- Unreasonable time constraints
- Lack of commitment from higher ups
- Not being responsible for the full scope
- Indecision on the part of the boss
- Increasing amounts of unfinished work
- Too much travel
- Lack of adequate project management tools
- Poor functional communications/writing skills
- Departmental "buck-passing"
- Meetings with executives
- Inability to relate to peers in a personal way
- Rush into decisions/beat the deadlines
- People being overpaid for their work
- Lack of reward ("pat on the back can do wonders")
- Expect too much from one's people and oneself
- Multiple time constraints
- Nonsupportive family
- Company political power struggles

- Need to get involved in details to get job done
- Not enough proven or trustworthy managers
- Vague goals and objectives
- Lack of a job description
- Too many people involved in minor decision making
- Lack of technical knowledge
- Disorganization of superiors
- No communication between sales and engineering
- Too much work for one person to handle effectively
- Excessive paperwork
- Lack of clerical/administrative support
- Work load growing faster than capacity
- Dealing with unreliable subcontractors
- Reeducating project managers
- Lack of new business
- Personnel not willing to take risks
- Demand for short-term results
- Lack of long-range planning
- Being overdirected
- Changing company systems that require relearning
- Overreacting management
- Poor lead time on projects
- Going from crisis to crisis
- Conflicting directives from above
- Line management acting as "father" figure
- Fire drills
- Lack of privacy
- Lack of challenge in job duties
- P.M. not involved/unknowledgeable about decision making
- Bureaucratic roadblocks ("ego")
- Empire-building line managers
- Disregard for company or personal things
- Documentation (reports/red tape)
- Large number of projects
- Inadequate or inappropriate requirements
- Desire for perfection
- Lack of dedication by technical experts
- Poor salary compared to contemporaries
- Lack of project organization
- Constant pressure
- Constant interruptions
- Problems coming in waves
- Severe home constraints
- Project monetary problems
- Shifting of functional personnel
- Lack of employee discipline
- Lack of qualified manpower

Sometimes the project manager's inability to effectively handle a time robber will create additional time robbers. Consider the following list of "How Not To Get Something Done:"[1]

- Profess not to have the answer. That lets you out of having any answer.
- Say that we must not move too rapidly. That avoids the necessity of getting started.

[1] *Source*: Dr. James Savage, Howard University, Washington, D.C.

- For every proposal, set up an opposite and conclude that the middle ground (no motion whatever) represents the wisest course of action.
- When in a tight place, say something that the group cannot understand.
- Look slightly embarrassed when the problem is brought up. Hint that it is in bad taste, or too elementary for mature consideration, or that any discussion of it is likely to be misinterpreted by outsiders.
- Say that the problem cannot be separated from other problems. Therefore, no problem can be solved until all other problems have been solved.
- Point out that those who see the problem do so because they are unhappy—rather than vice versa.
- Ask what is meant by the question. When it is sufficiently clarified, there will be no time left for the answer.
- Move away from the problem into endless discussion of various ways to study it.
- Put off recommendations until every related problem has been definitely settled by scientific research.
- Carry the problem into other fields; show that it exists everywhere; hence everyone will just have to live with it.
- Introduce analogies and discuss them rather than the problem.
- Explain and clarify over and over again what you have already said.
- As soon as any proposal is made, say that you have been doing it for 10 years.
- Wait until some expert can be consulted.
- Say, "That is not on the agenda; we'll take it up later." This may be extended ad infinitum.
- Conclude that we have all clarified our thinking on the problem, even though no one has thought of any way to solve it.
- Point out that some of the greatest minds have struggled with this problem, implying that it does us credit to have even thought of it.

EFFECTIVE TIME MANAGEMENT

To get more time out of each day, effective time management principles should be employed. The following list identifies such principles:

- Prepare a "to do" list and prioritize the activities. The list should be written down. Do not try to keep it in your head. Have one and only one list.
- Try to schedule uncommitted blocks of time each day to do the unexpected.

- Try to schedule daily committed blocks of time for the "must do" activities.
- As you begin each new task, ask yourself: Is this the best use of my time? Can this task be done equally well by someone else?
- Know the weekly and daily energy cycle of your people as well as your own. Be sure to assign or perform work that is compatible with this energy cycle.
- Make effective use of your secretary.
- If you have employees that come and go on flex-time schedules, be sure to account for this arrangement in assigning work and understanding their energy cycle.
- Meet with visitors in the reception area or any location other than your office.
- Understand the productivity level of your people and make sure that the project's performance standards are compatible with the productivity level.
- Do not schedule overtime unnecessarily unless you know that overtime is needed and that efficiency will be maintained. It is possible for employees to "save themselves" for overtime and thereby produce the same work in twelve hours that they would normally produce in eight hours.
- Refuse to do the unimportant or low-priority work.
- Avoid procrastination and start on the *most* difficult tasks first.
- Start now and look for ways to buy additional time.
- Be prepared to make quick decisions.
- Follow your schedule closely, especially critical items. You may find it necessary to monitor critical items yourself rather than wait for periodic feedback.
- Do not schedule meetings unless they are necessary.
- Assist your people in preparation for the meetings. Prepare an agenda and make sure that key personnel are informed well in advance of any major problems to be discussed.
- Conduct the meeting effectively and efficiently. Start the meeting on time, get right to the point, and end the meeting on time. Try to get all attendees to express their views and avoid prolonged discussions of trivial tasks. Do *not* discuss one-on-one problems in team meetings.
- Decide whether or not it is absolutely necessary for you to attend a given meeting, especially if travel is required.
- If travel is required, be prepared to work at travel stops.
- Be willing to delegate and employ the management-by-exception concept.
- Learn how to say no.

- Control telephone time and be willing to let your secretary take messages.
- Establish proper priorities for yourself, your project, and your team.
- Avoid writing memos. If memos or letters are necessary, make them short and summary-type in nature.
- If you have lengthy reports to read, it is best to take them with you on long trips, if time permits.
- Train your boss, peers, and subordinates on how to work with you.
- Be willing to assert your rights.
- If conflict resolution is necessary, obtain the required information as fast as possible and make a decision. Establishing procedures for conflict resolution early on in the project may be helpful.
- You must be willing to delegate to peers and subordinates. Don't be a "nice guy" and do it all yourself even though you may be able to do it quicker or better. Use the functional experts.
- Do not wait for someone else to make a decision that you can make.
- Be sure to "walk the halls" occasionally to find out what's going on in your project.
- Worry more about results than methods.
- Recognize that superiors are there to help, not to hinder you.
- Set goals and establish realistic deadlines.
- Group outgoing telephone calls at once or twice a day and keep them brief.
- Always keep a pocket notebook handy for writing down ideas.
- Handle each piece of paper once and only once.
- Save the routine or clerical activities for the low period in your energy cycle.
- Try to communicate without paperwork.
- Maintain one and only one file cabinet. (The wastebasket is there to be used.)

Project managers and project engineers must understand that even though they may have the authority, responsibility, and accountability for a project, they are still attached administratively to the parent organization and may have to cope with an additional set of time robbers and time management rules.

TIME MANAGEMENT FORMS

There are two basic forms which project managers and project engineers can use for practicing better time management. The first form is the "to do" pad as shown in Figure 8-1. The project manager or secretary pre-

Date ______________

Activities	Priority	Started	In Process	Completed

Figure 8-1. "To do" pad.

Date ______________

Time	Activity	Priority
8:00-9:00		
9:00-10:00		
10:00-11:00		
11:00-12:00		
12:00-1:00		
1:00-2:00		
2:00-3:00		
3:00-4:00		
4:00-5:00		

Figure 8-2. Daily calendar log.

pares the list of things to do. The project manager then decides which activities he must perform himself and assigns the appropriate priorities.

The activities with the highest priorities are then transferred to the "daily calender log" as shown in Figure 8-2. The project manager assigns these activities to the appropriate time blocks based upon his own energy cycle. Unfilled time blocks are then used for unexpected crises or for lower-priority activities.

If there are more priority elements than time slots, the project manager may try to schedule well in advance. This is normally *not* a good practice because it is very easy to create a backlog of high-priority activities to such a degree that schedule slippages are inevitable. In addition, an activity which today may be a "B" priority could easily become an "A" priority in a day or two. The moral here is *do not postpone until tomorrow what you or your team can do today.*

INTRODUCTION TO STRESS AND BURNOUT[2]

Everyone who works knows that on-the-job pressure is one of the major sources of stress in daily life. Project managers are subject to stress due to several different facets of their jobs, and this will be covered in more detail later in this section.

Stress is not necessarily negative. Without certain amounts of stress, reports would never get written or distributed, time deadlines would never be met, and, in fact, no one would ever even get to work on time. Stress serves not only to energize and motivate, but can also be a powerful force resulting in illness and even fatal diseases, and must be understood and managed if it is to be controlled and utilized for constructive purposes.

The mind, body, and emotions are not the separate entities they were once thought to be. They are one integrated system. One affects the other, sometimes in a positive way, and sometimes in a negative way. Stress becomes detrimental only when it is prolonged beyond what an individual can comfortably handle. In a project environment, with continuously changing requirements, impossible deadlines, and each project being considered as a unique entity in itself, we ask, How much prolonged stress can a project manager handle comfortably?

---

[2] The remaining sections have been adapted from Mary Khosh and Harold Kerzner, "Stress and Burnout in Project Management," presented at the Annual Seminar/Symposium on Project Management, sponsored by the Project Management Institute, Philadelphia, October 8–10, 1984.

Business people deal with these stresses in different ways. It is not unusual to find high-powered, successful executives "dropping out" and buying farms in Vermont. Nor is it unusual to find a project manager turning bicycle shop owner or house painter. When questioned, they will often say they did it because the pressures of their old jobs "weren't worth it."

Others are opting for early retirement at age 55 rather than continue to face the pressures of a demanding job. They may have successfully moved in their career to having responsibility for large projects involving millions of dollars and interfacing with all kinds of people. However, by then they might prefer not to take on another vast project. They often plateau out or develop a neurotic suspicion that every subordinate is competing for their job. In project management, peers may become subordinates. Responsibility increases threefold. Project managers may be caught in the vise of conflicting demands: demands from above to get more done with fewer people, and demands to work harder and longer to meet time constraints.

UNDERSTANDING STRESS

It is not unusual to find the "Peter principle" as the cause of stress when a person is promoted to a position which is beyond his capabilities. An engineer may be promoted to a position of project manager because of his previous strong technical performance, but when promoted, finds that interpersonal and communicative skills are also required to get the job done, and he may be poorly prepared for this part of his new assignment. As the work assignment becomes too demanding, insecurities in his talents and flaws in his self-image develop, and this lack of confidence further undermines his ability to do his job, which increases stress.

It is only the negative emotions which are bottled up inside that present health problems. The stress that harms is the stress that comes from fear or frustration, from continued anxiety or from persistent anger. When these emotions are continuous, they may create a multitude of medical problems.

Pressure, regardless of the source, forces a person to quicken the pace of his life style. It may cause him to intensify his dedication to a career or a project completion, or it may even motivate him to change goals.

Every individual has his own optimal stress level. He obtains the key to top performance when he determines what that level is. Successful people thrive on a certain amount of stress. It provides them with that extra drive which stimulates motivation and achievement. When one works at optimal stress levels, one is aware of an exhilaration not experienced if the pressures were absent. If one goes beyond that level, less productivity results, but when below that level one won't produce at peak performance levels or even experience the full potential for satisfaction in accomplishment. In fact, if one exists too far below his optimal stress level, he will actually vegitate.

Intense concentration produces changes in brain waves. In response to challenging or novel problems, the brain produces additional high-frequency brain waves. Ordinarily, beta brain waves, at 14–21 cycles per second, are used to get most jobs done. When solving more complex problems, gamma waves of 40 cycles per second begin to materialize. This focusing of attention on one task is necessary if the brain is to process information adequately. In project managers, the left hemisphere of the brain is predominant because it is the hemisphere which deals with logical, analytical, verbal, and mathematical abilities. It operates in a linear, sequential fashion. The left brain produces only one piece of information at a time. When too much stress or stimulus is presented at once, the higher cortical functions of the brain cannot operate properly and the ability to think and reason are consequently reduced. This is why a person under stress will tend to make more mistakes and errors in judgment than would be normal for him under conditions of no stress overload.

Optimal stress level is characterized by relaxed control under pressure, an optimistic outlook, heightened perceptual awareness, and a surge of energy.

STRESS IN PROJECT MANAGEMENT

The factors that serve to make any occupation especially stressful are responsibility without authority or ability to exert control, necessity for perfection, the pressure of deadlines, role ambiguity, role conflict, role overload, the crossing of organizational boundaries, responsibility for people, and the necessity to keep up with the information explosions or technological breakthroughs. Project managers have all of these factors included in their jobs.

A project manager has his resources controlled by line management, yet the responsibilities of bringing a project to completion by a prescribed deadline are his. A project manager may be told to increase the work output while the work force is simultaneously being cut. Project managers are expected to get work out on schedule but are often not permitted to pay overtime. One project manager described it this way: "I have to implement plans I didn't design, but if the project fails, I'm responsible."

Project managers, unlike line managers or top executives, do not have the power or facilities to accomplish many of their objectives alone. They must depend on superiors, subordinates, and peers for the cooperation and efforts that make their projects successful as they are constantly crossing organizational boundaries. Maintaining these three levels of interpersonal relationships is a juggling game that may make a consistent pattern of behavior almost impossible.

The project manager's superior is interested only in what is accomplished and may not be anxious to hear the specifics of how it was accomplished. The executive will hand down a list of necessary results, but it is the project manager's responsibility to translate that list into an organized system of behaviors that will produce the desired results. Whereas some levels of management have accessibility to a scapegoat if desired performance is not reached, the project manager has to assume full responsibility. He is evaluated on the results of the total operation and is solely responsible for whatever happens.

Additional high-intensity stressors in project management include unrealistic and inflexible time, cost and performance constraints, unrealistic customer and environmental constraints, no direct input into the staffing process, no direct control or authority over resources after project begins, no control over employee's purse strings, and having to share key personnel with other projects.

The stresses of project management may seem excessive for whatever rewards the position may offer. However, the project manager who is aware of the stresses inherent in the job and knows stress management techniques can face this challenge objectively and make it a rewarding experience.

THE JOB BURNOUT QUESTIONNAIRE

In order to grasp a better understanding of job stress and burnout in project management, a questionnaire was administered to 496 participants in three countries. The questionnaire was given in 1981–1983 to selected participants who attended training programs in basic, advanced, and engineering project management.

The 496 participants included 113 project managers, 124 project engineers, and 259 functional employees and functional managers. If an employee was a full-time project manager, with no other responsibilities, then he was categorized as a project manager. However, if he was a functional employee or line manager and was also acting as a project manager (i.e., wearing two hats), then he was listed in the "other" category.

In addition to present title, the questionnaire also asked length of time with company, length of time in present position; time span of projects;

Table 8-1. Industry Breakdown.

INDUSTRY	NUMBER OF RESPONDEES	AVERAGE AGE	AVERAGE PROJECTS RESPONSIBLE FOR AT SAME TIME	AVERAGE PROJECT SIZE
Oil/coal/mining	PM: 6	37	5	14M
	PE: 5	41		
	Other: 6	37		
Manufacturing	PM: 19	35	4	4M
	PE: 37	36		
	Other: 91	38		
Computers	PM: 10	36	4	5M
	PE: 4	33		
	Other: 17	34		
Government	PM: 4	38	5	3M
	PE: 6	38		
	Other: 7	39		
Chemical	PM: 3	45	10	12M
	PE: 1	25		
	Other: 5	45		
Automotive	PM: 3	42	10	177M
	PE: 2	25		
	Other: 6	38		
Paper	PM: 0	—	—	—
	PE: 5	37		
	Other: 3	45		
Utilities	PM: 5	43	7	1B
	PE: 10	38		
	Other: 21			
Construction	PM: 6	40	3	8M
	PE: 3	35		
	Other: 7			
Engineering	PM: 9	41	4	61M
	PE: 8	43		
	Other: 14	39		
Railroad	PM: 1	35	8	5M
	PE: 0	—		
	Other: 0	—		

Table 8-1. (continued).

INDUSTRY	NUMBER OF RESPONDEES	AVERAGE AGE	AVERAGE PROJECTS RESPONSIBLE FOR AT SAME TIME	AVERAGE PROJECT SIZE
Tobacco	PM: 3	32	3	5M
	PE: 0	—		
	Other: 0	—		
Repair/maint.	PM: 0	—	—	—
	PE: 0	—		
	Other: 2	55		
Industry unkn.	PM: 5	33	3	13M
	PE: 0	—		
	Other: 1	35		
International	PM: 10	39	3	12M
	PE: 19	37		
	Other: 25	41		

dollar value of projects; number of projects responsible for at one time; and amount of overtime required.

Partial results are tabulated per industry as shown in Table 8-1. Almost all responders had been with their present company between five and ten years, in their present position for two to five years, and working an average of ten hours of overtime per week.

The intent of the questionnaire was to survey both junior and senior project managers in both large and small companies.

THE PROJECT MANAGEMENT STRESS MODEL

There are four regions (identified as I, II, III, and IV) and four interface zones (A, B, C, and D), in the stress model which are areas common to more than one region. This is shown in Figure 8-3.

Region I, the time management region, may very well be the most important stressor. Project managers enter this region with hopes and aspirations to perform according to plan. Unfortunately, if the project managers are unable to identify and control the time robbers, severe stress can result.

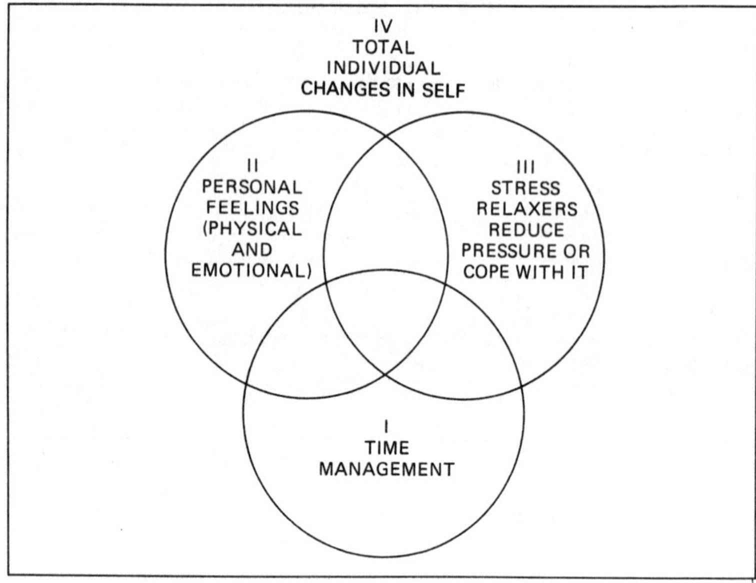

Figure 8-3. Personal feeling (job-related or otherwise) time management on job.

The questionnaire asked the participants to rank the following time robbers and hassles which appear in region I:

- Telephone interruptions
- Unscheduled visitors or "drop-ins"
- Meetings
- Having to do jobs over
- Paperwork
- Delayed decision
- Casual conversations
- Lack of sufficient clerical support
- Disorganization of co-workers
- Confused responsibility and authority
- Physical work environment
- Disorganization of superiors
- Poor communications between employees
- Having to explain thinking to superiors
- Superiors' failure to delegate
- Crises
- Lack of organizational goals and objectives
- Changes in corporate plans, rules, and regulations

- Poor communications with superiors
- Lack of authority to delegate
- Incomplete work
- Waiting for people
- Red tape
- Poor retrieval systems
- Performing routine tasks
- Unclear priorities set by superiors

Region II, the "emotions" region, includes the individuals' personal feelings (i.e., physical and emotional) resulting from hassles and the time management problems. The questionnaire asked the participants to evaluate the following emotions and feelings:

- Being tired
- Feeling depressed
- Having a good day
- Being physically exhausted
- Being emotionally exhausted
- Being happy
- Feeling "wiped out"
- Feeling "burned out"
- Being unhappy
- Feeling rundown
- Feeling trapped
- Feeling worthless
- Feeling weary
- Being troubled
- Feeling disillusioned and resentful about people
- Feeling weak and helpless
- Feeling hopeless
- Feeling rejected
- Feeling optimistic
- Feeling energetic
- Feeling anxious

Obviously, these feelings can be influenced by home life, personal problems, etc., rather than job-related hassles and time management. The people filling out the questionnaire were instructed to consider only job-related causes and effects.

Region III is where the project manager realizes that he must cope with stress and tries to reduce or cope with the pressure by using stress relaxers. The project managers/other personnel were asked to rate each of the following:

- I work through lunch.
- I exercise regularly.
- I have at least one alcoholic drink each day.
- I take work home on weekends.
- I receive adequate rewards, appreciation, and recognition from my boss.
- I receive strong support from my family or close friends.
- I perform a regular method of relaxation.
- I share my problems with a friend or spouse.
- My boss and I discuss my work problems.

Region IV is where the project manager analyzes the changes in himself over the past six months as a result of regions I, II, and III. The intent here was to force the participant to recognize the changes in himself as a result of his ability or inability to handle stress. The students answered the following questions:

- Do you tire more easily? Do you feel fatigued rather than energetic?
- Are people annoying you lately by telling you "You don't look so good?"
- Are you working harder and harder and accomplishing less and less?
- Are you often invaded by a sadness that you cannot explain?
- Are you forgetting appointments, deadlines, personal possessions, etc.?
- Are you increasingly irritable? More short-tempered? More disappointed in the people around you?
- Are you seeing close friends and family members less frequently?
- Are you too busy to do even routine things like phone calls or read reports?
- Are you suffering from physical complaints (aches, pains, and lingering cold)?
- Do you feel disenchanted when the activity of the day comes to a halt?
- Is joy elusive?
- Are you unable to laugh at a joke about yourself?
- Does sex seem more trouble than it's worth?
- Do you have very little to say to people?

In addition to the four regions, there are also four interface zones; A, B, C, and D. Zone A is a transition or interface zone between hassles, time management, and personal feelings and emotions. In this transition zone, the hassles and other time management problems create stressors which affect the employee's emotions, physical performance, and health.

Transition zone A exists continuously throughout the life of the project. The intensity of the forces that exist in the zone depend upon one's ability to handle the continuous stream of stressors. Good time management practices, either through administrative policies and procedures or through effective delegation, reduce the size of this zone.

This zone is in existence at the birth of all projects and may grow rather than shrink in size. This zone becomes extremely turbulent for project managers responsible for multiple projects, each within a different life-cycle phase.

Zone B is the transition zone between personal feelings and stress relaxers. In this zone, the individual realizes that he is under stress and attempts to find relaxers.

Zone C is the transition zone between stress inducers and stress relaxers. The size and intensity of this zone is usually dependent upon the experience and maturity of the project manager. Experienced project managers can quickly size up tasks and activities which are stress producers and know how to take immediate corrective action.

Obviously, the ideal situation is for an individual to be experienced enough such that he continuously performs in zone C. Unfortunately, in the real world, the uniqueness of each project and each task within each project forces continuous reassessments and thus most people perform in transition zone D, which is simply the interfacing of all three transition zones.

It should be obvious that the model is a management-type model rather than a psychological-type model. The intent here is to show that stress, hassles, and time robbers can be managed in a project environment.

In any given model, the validity rests with the assumptions. The major assumption made here is that all stressors are job-related and that the questionnaire does not contain any bias from one's personal life.

REGION I: TIME MANAGEMENT

The majority of all stress in project management is deeply rooted in time management pitfalls and problems. The younger project managers normally face the biggest time management challenges because they often feel that they must perform the job themselves, whereas the more experienced project managers delegate more often.

Table 8-2. Ranking of Major Time Robbers.

RANK	PM	PE	OTHERS
1	Telephone	Paperwork	Crises[b]
2	Delayed decisions[a]	Delayed decisions	Delayed decisions[b]
3	Unclear priorities[a]	Waiting for people	Paperwork
4	Paperwork	Crises	Meetings
5	Red tape	Red tape	Employee comm.
6	Resp./authority	Disorg. superiors	Red tape

[a] Tied for 2/3.
[b] Tied for 1/2.

In some industries (primarily project-driven), project management *cannot* be accomplished in less than 50 hours per week, with the project manager's time subdivided as

- Meetings: 10 hours/week
- Conflict resolution: 8 hours/week
- Planning/replanning: 10 hours/week
- Reading/writing: 12 hours/week
- Communication with colleagues, superiors, etc.: 10 hours/week

The fortunate project managers can get by with 40 hours per week.

During the early stages of evolution of project management, project managers generally spend more time "doing" rather than managing. This is a result of the project manager's desire not to delegate because he can obviously do it better than anyone else. As project management matures, project managers delegate more and more, even portions of their administrative responsibilities. It is a false assumption that only the administrative time robbers affect the project managers and the operational time robbers affect the project engineers and line personnel.

Table 8-3. Ranking of Minor Time Robbers.

RANK	PM	PE	OTHERS
1	Redo jobs	Redo jobs	Work environment
2	Work environment	Work environment	Authority/resp.
3	Comm. w/superiors	Casual convers.	Redo jobs
4	Explain to superiors	Explain to superiors	Comm. w/superiors
5	Policy changes	Comm. w/superiors	Explain to superiors
6	Casual convers.	Retrieval system	Policy changes

Table 8-4. Industries with Greatest Time Robbers.

RANK	PM	PE	OTHERS
1	Nuclear	Medical	Oil/coal
2	Services	Services	Metals
3	Manufacturing	Nuclear[a]	International
4	Railroad	Metals[a]	Services
5	Metals	Automotive[b]	Nuclear
6	Elect./aerospace	International[b]	Automotive

[a] Tied for 3/4.
[b] Tied for 5/6.

The results shown in Tables 8-2 and 8-3 identify the commonality of the major time robbers, indicating that the majority of the participants were in companies that had either mature project management or were approaching it. As expected, the intensity of the time robbers and hassles were greater for PEs/others, indicating that the project managers may be delegating more.

The commonality of the low-intensity time robbers indicate those of a mature project management organization where employees have learned how to continuously cope with adversity.

Other data collected (as shown in Tables 8-4 and 8-5) indicate those industries which identify the greatest and least intensity of time robbers and hassles. In the service, nuclear, and metal industries, there appears to be agreement between the PMs, PEs, and functional personnel. But in the oil/coal, electronics/aerospace, automotive, and manufacturing industries, there appear to be major differences as to the intensity of the time robbers.

Table 8-5. Industries with Least Time Robbers.

RANK	PM	PE	OTHERS
1	Education	Computers	Unknown indust.
2	Automotive	Construction[a]	R&D
3	Chemical	Military[a]	Elec./aerospace
4	Government	Oil/coal	Computers
5	Construction	Manufacturing	Engineering
6	Tobacco	Paper	Government

[a] Tied for 2/3.

REGION II: THE "EMOTIONS" REGION

The physical and emotional feelings that one must recognize and endure are usually combinations of personal, home-life, and job-related experiences. For simplicity's sake, we will analyze these feelings simply from a job-related experience.

The questionnaire identified 21 feelings.

1. *Being tired:* Being tired is a result of being drained of strength and energy, perhaps through physical exertion, boredom, or impatience. The definition here applies more toward a short-term rather than long-term effect. Typical causes for feeling tired include meetings, report-writing, and other forms of document preparation.

2. *Feeling depressed:* Feeling depressed is an emotional condition usually characterized by discouragement of a feeling of inadequacy. There are several sources of depression in a project environment: Management or the client considers your report unacceptable, you are unable to get timely resources assigned, the technology is not available, and the constraints of the project are unrealistic and may not be met.

A state of depression in a project environment is usually the result of a situation which is beyond the control or capabilities of the project manager. This situation can exist indefinitely because the project manager has a great deal of responsibility and very little authority, and has no direct control over the staffing or assignment of personnel.

3. *Having a good day:* For most people, having a good day implies that something has gone right. In a project management environment, every project may be inherently different and a happy day may simply be the ability to attack and resolve difficult problems, even if the final result is detrimental to the project. Sometimes it is truly amazing how decision making and looking at the "means to the end" rather than the end itself equals to a good day. An experienced project manager once commented: "If I sit at my desk and do not have problems to resolve, then I go out and look (even create) problems to solve. That's my definition of having a good day."

4. *Being physically exhausted:* Project managers are both managers and doers. It is quite common for project managers to perform a great deal of the work themselves because either they consider the assigned personnel unqualified to perform the work or because the project manager is impatient and considers himself capable of performing the work faster. In addition, project managers often work a great deal of "self-inflicted" overtime.

5. *Being emotionally exhausted:* The most common cause of emotional exhaustion is report writing and the preparation of handouts for inter-

change meetings. Sometimes the project manager finds himself performing these functions for line personnel, but more often than not, line employees procrastinate and force this function upon project managers. Since data preparation is a continuous project function, one might expect this effect to occur frequently.

6. *Being happy:* Happiness generally suggests a feeling of pleasure and contentment. Most project managers view project management as a lifetime profession and are usually quite happy even under situations of stress. A senior construction project manager commented on why he has not accepted a promotion to vice-president: "I can take my children and grandchildren into 10 countries in the world and show them projects which I either built or help build. What do I show them as a vice-president? My bank account? The size of my office? The stockholder's report?" Obviously not all people would respond like this. The work challenge associated with the project environment is a strong driving force toward happiness. There are very few positions where an employee can see an activity through from beginning to end.

7. *Wiped out:* Feeling wiped out is normally a combination of physical and emotional exhaustion. It is a short-term rather than long-term effect and may be caused by short spurts of intense overtime; nearness of deadlines on the time, cost, and performance constraints; or simply lengthy customer review meetings.

8. *Burned out:* Being burned out is more than just a feeling; it is a condition. Being burned out implies that one is totally exhausted, both physically and emotionally, and that rest, recuperation, or vacation time may not remedy the situation. The most common cause is prolonged overtime, or the need thereof, and an inability to endure or perform under continuous pressure and stress. The solution is almost always a change in job assignment, preferably with another company.

Burnout can occur almost overnight, often with very little warning. In one company, a project manager who was managing high-technology projects (and was on the fast track in the organization) burned out and accepted a job as the manager of quality control for a small company that manufactures brooms. In the termination interview, the project manager stated that his reason for leaving was because he felt burned out.

9. *Being unhappy:* There are several factors which produce unhappiness in project management. Such factors include highly optimistic planning, unreasonable expectations of management, management cutting resources because of a "buy-in," or simply customer demands for additional data items. A major source of unhappiness is the frustration caused by having limited authority which is not commensurate with the assigned responsibility.

10. *Feeling run down:* Feeling run down is a temporary condition caused by exhaustion, overwork, or simply poor physical conditioning. The run-down feeling usually occurs following "panics," especially as one nears the project constraints.

11. *Feeling trapped:* The most common situation where project managers feel trapped is when they have no control over the assigned resources on the project and feel as though they are at the mercy of the line managers. Employees tend to favor that manager who can offer them the most rewards, and that is usually the line manager. Providing the project manager with some type of direct and indirect reward power can remedy the situation.

Another trapped feeling is when the project manager and line managers work together to develop realistic costs and schedules for a proposal and senior management arbitrarily slashes the price to remain competitive. Now, if the project is awarded to the company, the project manager feels trapped into accepting constraints which really are not his.

12. *Feeling worthless:* Feeling worthless implies without worth or merit, that is, valueless. This situation occurs when project managers feel that they are managing projects beneath their dignity. Most project managers look forward to the death of their project right from the onset, and expect their next project to be more important, perhaps twice the cost, and more complex. Unfortunately, there are always situations where one must take a step backwards.

13. *Being weary:* Being weary is a combination of several of the previous feelings. This includes tiredness, lack of energy, worn out, and perhaps impatient. Weariness is usually the transition stage between being tired and burned out.

14. *Being troubled:* The two major causes for a project manager to be troubled are optimistic planning (which has since turned "sour") and approaching the constraints of the project with little hope for correction. The latter case is not a really serious problem because most project managers pride themselves on their ability to perform trade-offs.

15. *Feeling disillusioned and resentful about people:* This situation occurs most frequently in the project manager's dealings (i.e., negotiations) with the line managers. During the planning stage of a project, line managers often make promises concerning future resource commitments but renege on their promises during execution. Disillusionment occurs and can easily develop into serious conflict. Another potential source is when line managers do not appear to be making decisions which are in the best interest of the project.

16. *Feeling weak and helpless:* The weak and helpless feeling is a common result from feeling disillusioned. Again, the cause of this feeling

depends upon the working relationship that project managers have with executives and line managers.

17. *Feeling hopeless:* The most common sources of hopelessness are R&D projects where the ultimate objective is beyond the reach of the employee or even the state-of-the-art technology. Hopeless means showing no signs of a favorable outcome. Hopelessness is more a result of the performance constraint than time or cost.

18. *Feeling rejected:* Feeling rejected can be the result of a poor working relationship with executives, line managers, or clients. Rejection often occurs when people with authority feel that their options or opinions are better than those of the project manager. Rejection has a demoralizing effect upon the project manager because he feels that he is the "president" of the project and the true "champion" of the company.

19. *Feeling optimistic:* Almost all project managers feel optimistic, even in time of trouble. Project managers often have more faith in themselves and others than other people perceive. Optimism is usually a desired trait in a project manager.

20. *Feeling energetic:* The work challenge created by the project environment usually brings with it an energetic feeling, where the individual accepts the daily challenge of problem solving and troubleshooting. The exact degree of energy may depend, of course, upon the time of day, the day of the week, or simply the age of the project manager.

21. *Feeling anxious:* Almost all project managers have some degree of "tunnel vision" where they look forward to the end of the project, even when the project is in the infancy stages. This anxious feeling is not only to see the project end, but to see it completed successfully. The anxiousness can also be applied to show one's ability to solve complex problems.

CATEGORIZING THE FEELINGS

The first step in the analysis of the feelings was to see if there were direct correlations between the feelings and the ages and years of service of the project managers, project engineers, and other employees. The conclusions are as follows:

1. Almost all of the project managers, project engineers, and other employees were approximately the same age for each feeling and with the same years of experience. This was expected, since most of the questionnaires were filled out by seminar participants who were relative "novices" to project management.
2. For project managers, the average age was younger for those feeling tired, wiped out, burned out, unhappy, run down, trapped, and

worthless. This implies that either severe stress and pressure are being placed upon the younger project managers, or else employees are being placed into project management with very little training. The latter appears more probable.

The feelings can be further categorized into three groupings:

1. Positive feelings (category I feelings)
   - Having a good day
   - Happy
   - Optimistic
   - Energetic
   - Anxious
2. Negative feelings over which the project manager may have some control (category II feelings)
   - Tired
   - Physical exhaustion (*)
   - Emotional exhaustion (*)
   - Wiped out
   - Weary (*)
   - Run down
   - Unhappy
3. Negative feelings over which the project manager may not have direct control (category III)
   - Depressed
   - Physical exhaustion (*)
   - Emotional exhaustion (*)
   - Burned out
   - Trapped
   - Worthless
   - Weary (*)
   - Troubled
   - Disillusioned
   - Weak/helpless
   - Hopeless
   - Rejected

The feelings indicated by an asterisk (*) are considered common to both categories II and III. Actually, all of the feelings can be regarded as interrelated, but will be separated here for simplicity's sake.

The categories of feelings can now be used for further investigations. The feelings that can result from the project manager's working relationship with the project office, superiors, and line departments are shown in Table 8-6. As expected, category I feelings can exist with the project office, superiors, or line departments if the project appears to be heading in the right direction and with minimum deviations from time, cost, and performance.

Since the project manager may have direct control over only the project office, it is reasonable to assume that category II feelings occur in the project manager/project office working relationship. Similarly, the category III feelings over which the project manager may have no control occur in dealings with superiors (and possibly customers).

The categories of feelings can also be related to the life-cycle phases of a project as shown in Table 8-7. This assumes that all projects can be broken down into four phases. In the conceptual phase, the project manager's energy level and enthusiasm are generally high. Here, the project manager may have minimum support from the line groups and must rely upon his own ability and ingenuities. Work challenge and desire to achieve will be at a maximum, together with the project manager's energy level.

Table 8-6.

CATEGORY OF FEELINGS	PM WORKS WITH		
	PROJECT OFFICE	SUPERIORS	LINE DEPARTMENTS
Tired			X
Depressed		X	X
Having a good day	X	X	X
Physical exhaustion	X		X
Emotional exhaustion	X		X
Happy	X	X	X
Wiped out	X		X
Burned out	X		X
Unhappy		X	X
Run down	X		
Trapped		X	X
Worthless		X	
Weary	X		
Troubled		X	X
Disillusioned		X	X
Weak/helpless		X	
Hopeless		X	X
Rejected		X	X
Optimistic	X	X	X
Energetic	X	X	X
Anxious	X	X	X

Table 8-7.

CATEGORY OF FEELINGS	LIFE-CYCLE PHASE			
	CONCEPTION	PLANNING	MAIN	PHASEOUT
Tired			X	X
Depressed		X	X	
Having a good day	X	X	X	X
Physical exhaustion			X	X
Emotional exhaustion			X	X
Happy	X	X	X	X
Wiped out			X	X
Burned out			X	X
Unhappy		X	X	X
Run down			X	X
Trapped		X	X	
Worthless	X	X	X	X
Weary			X	X
Troubled		X	X	
Disillusioned		X	X	
Weak/helpless			X	
Hopeless			X	X
Rejected	X	X	X	X
Optimistic	X	X	X	
Energetic	X	X	X	
Anxious	X	X	X	

The planning phase is where the project manager has the opportunity to demonstrate his true abilities to successfully plan a project. Motivation, desire to achieve, work challenge, and energy are extremely high. The project manager feels that he is about to become a "star."

The main phase may very well be the most difficult for the project manager and can be either feast or famine. The difficulty with the main phase is that the project manager must observe the work being performed by resources that are beyond his control. Customers and executives demand results, and the line departments want more money (hours) or a change in project specifications.

In the last life-cycle phase, phaseout, the project manager must wind down and prepare for the closeout of the project. At this point, the project manager realizes that he is totally accountable for the success or failure, and may spend more time worrying about his next assignment rather than the successful termination of his present assignment. The project manager's energy level is normally low during phaseout.

In the conceptual phase, category I feelings dominate. In the planning phase, category I feelings are dominant but may be dependent upon the project manager's involvement with executive management and line management. If executive and line management "pull rank" on the project manager during the planning phase, then category III feelings will occur.

Although the planning phase is considered the most important with regard to project success, the main phase has the greatest impact on the feelings of the project manager. In the main phase, all three categories of feelings are possible and, depending upon the associated problems, one can easily transition back and forth between categories of feelings.

The phaseout period is usually associated with category III feelings. This holds true regardless of whether or not the project was successful. If the project is a failure, the project manager enters emotional depression (whether or not the results were beyond the project manager's control) and category III feelings dominate. If the project is successful, the project manager may be looking forward to his next assignment rather than completing his present responsibility. Both positive and negative wind-down can produce category III feelings.

The reader must remember that the category of feelings is dependent upon the hassles and *time robbers* within each phase, not necessarily upon the phase itself. As shown in Figure 8-4, category III feelings dominate for the majority of the hassles and time robbers. This is to be expected since these hassles and time robbers can occur frequently in the main phase and phaseout period of a project's life cycle.

The highest industry averages appeared for those organizations that were project-driven. This was to be expected. The service and metal industries had the greatest averages for project managers and project engineers.

An interesting result was that the project managers, project engineers, and functional personnel all selected the same top six feelings: namely, tired, having a good day, happy, optimistic, energetic, and anxious. This result implies the project personnel are becoming more knowledgeable in project management and are learning to cope well under situations of stress which are accompanied by a multitude of hassles and time robbers.

The feelings category was compared with project managers, project engineers, and "other" employees, for all industries combined. As anticipated, all three groups of personnel felt that their category I feelings were almost twice the frequency of the category III feelings. This is an extremely positive indication that

- Project managers and engineers are delegating more.
- Functional groups are well staffed with qualified personnel.
- Customers and executives are identifying more realistic milestones.
- Good policies and procedures have been developed for project planning, scheduling, and control.

	TIME ROBBERS																									
FEELING	TELEPHONE	VISITORS	MEETINGS	DO JOBS OVER	PAPERWORK	DELAYED DECISIONS	CASUAL CONVERSATIONS	POOR CLERICAL SUPPORT	DISORGANIZATION OF P.O.	CONFUSED AUTHOR/RESP.	ENVIRONMENT (PHYSICAL)	DISORG. OF SUPERIORS	COMM. BETWEEN EMPLOYEES	EXPLAIN TO SUPERIORS	SUPERIORS FAIL TO DELEGATE	CRISES	UNCLEAR GOALS	PLANNING CHANGES	LACK OF AUTHORITY	INCOMPLETE WORK	WAITING FOR PEOPLE	RED TAPE	POOR RETRIEVAL SYSTEMS	ROUTINE TASKS	UNCLEAR PRIORITIES	POOR COMM. WITH SUPERIORS
TIRED	X	X	X		X						X					X					X	X		X		
DEPRESSED			X		X				X	X	X		X		X	X	X		X	X	X	X	X		X	
HAVE A GOOD DAY											X		X	X		X		X								
PHYS. EXHAUST.		X	X		X						X					X						X				
EMOT. EXHAUST.			X		X						X			X		X						X				
HAPPY					X						X			X		X		X								
WIPED OUT			X								X					X						X				
BURNED OUT											X			X		X						X				
UNHAPPY		X	X	X	X	X		X	X	X	X	X	X	X	X	X	X			X	X	X	X	X	X	X
RUNDOWN			X		X					X	X					X						X				
TRAPPED		X	X		X			X	X	X	X	X	X	X	X				X		X	X		X		
WORTHLESS					X					X	X		X	X	X				X			X				
WEARY	X		X		X		X	X	X	X	X		X									X		X		
TROUBLED					X				X	X	X	X	X			X	X			X	X	X	X			
DISILLUSIONED						X		X	X	X	X	X		X	X				X		X	X				X
WEAK/HELPLESS					X	X		X	X	X	X	X	X	X	X	X			X		X	X	X	X	X	X
HOPELESS										X	X	X		X	X		X				X	X				
REJECTED						X			X	X	X	X		X	X				X		X	X			X	X
OPTIMISTIC					X						X		X	X		X		X								
ENERGETIC					X						X			X		X		X								
ANXIOUS					X						X			X		X		X								

Figure 8-4

Table 8-8 shows what percent of the hassles and time robbers were attributed to each feeling. For example, those PMs/PEs who considered themselves as rarely feeling tired (choices 1–3) accounted for 23.4% of the time robber intensities, whereas those who felt tired more often (choices 4–7) accounted for 76.6%. As before, the PMs/PEs who selected high frequencies for tired, having a good day, happy, optimistic, energetic, and anxious accounted for better than 75% of the time robbers. On the *low frequency* side, those individuals who selected burned out, worthless, and hopeless also correlated well.

One of the most interesting results was found when the feelings in the experience region were compared one against one. For example, 95% of

Table 8-8. Correlating Personal Feelings with Time Robbers (for PM/PE only).

FEELINGS	TIME ROBBERS (%)
Tired (1–3)	23.4
Tired (4–7)	76.6
Depressed (1–3)	57.3
Depressed (4–7)	42.7
Have a good day (1–3)	6.7
Have a good day (4–7)	93.3
Physical exhaustion (1–3)	64.5
Physical exhaustion (4–7)	35.5
Emotional exhaustion (1–3)	49.4
Emotional exhaustion (4–7)	50.6
Happy (1–3)	6.5
Happy (4–7)	93.5
Wiped out (1–3)	73.4
Wiped out (4–7)	26.6
Burned out (1–3)	81.1
Burned out (4–7)	18.9
Unhappy (1–3)	67.0
Unhappy (4–7)	33.0
Run down (1–3)	61.2
Run down (4–7)	38.8
Trapped (1–3)	60.1
Trapped (4–7)	39.9
Worthless (1–3)	89.0
Worthless (4–7)	11.0
Weary (1–3)	50.6
Weary (4–7)	49.4
Troubled (1–3)	36.7
Troubled (4–7)	63.3
Disillusioned (1–3)	58.0
Disillusioned (4–7)	42.0
Weak/helpless (1–3)	79.9
Weak/helpless (4–7)	20.1
Hopeless (1–3)	90.4
Hopeless (4–7)	9.6
Rejected (1–3)	84.6
Rejected (4–7)	15.4
Optimistic (1–3)	3.4
Optimistic (4–7)	96.6
Energetic (1–3)	4.1
Energetic (4–7)	95.9
Anxious (1–3)	23.2
Anxious (4–7)	76.8

the participants who selected "tired" also selected "having a good day." An interesting result is the high correlation of "optimistic," "energetic," and "anxious" with all of the other feelings. Functional employees/others identified a 75% correlation between "tired" and "emotionally exhausted," whereas the PM/PEs did not correlate, perhaps because they were more accustomed to these stresses.

REGION III: STRESS RELAXERS/COPING METHODS

If the feelings that one has produces stress, then the natural response is to seek coping methods or stress relaxers. The thickness of the interface zone between stress and stress relaxers depends not only upon the identification of stress, but selection, frequency, and magnitude of the coping methods or stress relaxers. The individuals surveyed were given nine choices for coping methods/stress relaxers. These choices included working through lunch; regular exercise; drinking through the day; taking work home on weekends; receiving strong support; receiving adequate rewards, appreciation, and recognition from boss; regular relaxation; sharing problems with family members and friends; discussing work problems with boss.

The results indicated that line employees, as well as project office personnel, work through lunch; less than 50% of the participants exercise regularly or employ relaxation methods; slightly more than 50% receive recognition (rewards); and 30% rely upon superiors to help resolve problems. The two unusual results were strong support provided by friends and contemporaries, and that only 20% regularly take work home on weekends.

REGION IV: RECOGNIZING CHANGES

Recognizing stress and taking corrective action can be effective only if one perceives that there may be detrimental changes taking place in one's personality, habits, life style, and home life. Without this type of recognition or feedback, the selection of coping methods or stress relaxers may provide only a short-term effect rather than long-term solutions.

The survey asked participants to select from 15 recognizable changes in themselves on a scale of 1–5, where 1 = little or no change and 5 = a great deal of change. The same ranking for changes was observed in both PM/PEs as well as functional personnel/others, indicating that the entire organization feels the impact of project management, not merely the project office personnel.

The results are tabulated in Table 8-9 showing the correlation between recognized changes in oneself and stress relaxers and time robbers or hassles. As expected, those individuals who perceived major changes in themselves had a higher correlation with coping methods/stress relaxers and hassles and time robbers. Yet the low-frequency correlations were substantially higher-correlated than the high-frequency responses. This implies that either the participants are adjusting to the situations, or they simply do not recognize the changes taking place.

Table 8-9. Comparing Recognized Changes in Oneself with Existing Time Robbers and Stress Relaxers (for PM/PE only).

RECOGNIZED CHANGES IN ONESELF	TIME ROBBERS (%)	STRESS RELAXERS (%)
Tire easily (1–2)	62.6	64.4
Tire easily (3–5)	37.4	35.6
Not looking good (1–2)	87.3	87.8
Not looking good (3–5)	12.7	12.2
Working harder (1–2)	76.8	75.6
Working harder (3–5)	23.2	24.4
Cynical and disenchanted (1–2)	68.2	67.7
Cynical and disenchanted (3–5)	31.8	32.3
Sadness you can't explain (1–2)	83.8	81.9
Sadness you can't explain (3–5)	16.2	18.1
Forgetful (1–2)	76.5	75.2
Forgetful (3–5)	23.5	24.8
Irritable (1–2)	66.9	66.8
Irritable (3–5)	33.1	33.2
Seeing friends and family less freq. (1–2)	69.0	68.4
Seeing friends and family less freq. (3–5)	31.0	31.6
Too busy to do routine (1–2)	60.3	58.8
Too busy to do routine (3–5)	39.7	41.2
Physical complaints (1–2)	86.4	86.8
Physical complaints (3–5)	13.5	13.2
Disoriented (1–2)	76.0	76.9
Disoriented (3–5)	24.0	23.1
Joy elusive (1–2)	81.0	79.7
Joy elusive (3–5)	19.0	20.3
Unable to laugh (1–2)	92.2	92.5
Unable to laugh (3–5)	7.8	7.5
Sex more trouble (1–2)	86.4	87.0
Sex more trouble (3–5)	13.6	13.0
Very little to say (1–2)	79.0	78.4
Very little to say (3–5)	21.0	21.6

Several characteristics have been identified through research as being attributes of individuals who are effective stress managers:

- Knowledge of self-strengths, skills, and liabilities
- Varied interests—many sources of satisfaction
- Variety of reactions to stress—repertoire of coping responses or stress relaxers
- Acknowledgment and acceptance of individual differences
- Active and productive—participate in vigorous regular exercise

SUMMARY

Time is a nonrenewable resource and should be used accordingly. In this chapter we discussed the idea of "time robbers" and how managers and professionals can better use their time.

Our discussion of stress and burnout was related to the stress factor in project management. Recognizing stress and seeking help, or taking action on your own, can be effective only if the individual perceives that stress can come from work and other environments in which he lives, works, or plays. Without this type of feedback or recognition, coping with strategies and methodologies may provide only short-term results.

Stress and/or burnout are real dangers facing managers and professionals in today's complex team-driven organizations. This chapter has given them a good start in coping with such dangers.

In Chapter 9, the concluding chapter of this book, the important subject of team building is discussed.

BIBLIOGRAPHY

Khosh, Mary and Harold Kerzner. "Stress and Burnout in Project Management." *Annual Seminar/Symposium on Project Management,* Project Management Institute, Philadelphia, October 8–10, 1984.

9
A FRAMEWORK FOR DEVELOPING HIGH-PERFORMING TECHNICAL TEAMS[1]

INTRODUCTION

A substantial contribution to the future success of American technology-based industry will come from various types of technical teams. Only through highly coordinated teamwork will it be possible to reach increasingly complex technological goals with the speed and flexibility required in the turbulent world marketplace. By *teams* we mean all sizes and kinds of groups of technical specialists who are for any reason interdependent for any length of time. Teams can be intact departments or divisions whose work is ongoing, project groups assembled to accomplish large-scale but finite tasks, or task forces and committees created for specific purposes. Team members may or may not have other duties outside the team.

Oftentimes the engineers, technical specialists, and/or scientists who make up a particular team are drawn from all over the organization and differ widely in specialties, backgrounds, and experience. Some may also have more status, formally or informally, than others, perhaps even more than the person designated as the team manager. Whoever *is* the team manager, then, may be facing more than just the work. He or she must be prepared to deal with divided loyalties and interpersonal conflict. There is generally a high correlation between a team's task performance and the "health" of the interpersonal relations among members. Poor interper-

[1] This chapter was written by Dr. Judith Mower and Dr. David Wilemon. Dr. Judith Mower is an organizational psychologist and cofounder of ManageMentors, a Syracuse-based management consulting firm. She has trained managers from numerous organizations in team leadership, conflict resolution, and other aspects of management competency. She is also an adjunct faculty member of the School of Management at Syracuse University. Dr. David Wilemon is Director of the Innovation Management Program at Syracuse University's School of Management. He has published widely in the areas of project management, team building, conflict management, and the management of technical organizations.

sonal relations can cause poor performance, and vice versa.[2] Effective team management, therefore, requires careful attention to *both* task performance and interpersonal relations.

FOCUS

We believe a simple framework for steering and motivating both the task and interpersonal progress of the team would be most useful to technical team managers. The framework presented here is called *team development*. The ideas that emanate from that framework are called actions for *team building*.

Team development is a process similar to individual development. Just as people mature through certain phases, teams go through noticeable phases. Teams mature in terms of task progress and in terms of interpersonal relations. At each phase of development certain problems can occur. Team managers can stave off such problems through team building. In this way team progress is not slowed. Here are some situations where team building could have made a difference:

> *Situation A:* A manager racing to get his team off to a quick start failed to discuss adequately the overall project management plan with the team. As a result, confusion developed over project direction and priorities. A great deal of "wheel-spinning" resulted, leading to intense frustration and apathy.
>
> *Situation B:* A team charged with developing a new product for surgical use was unable to agree on which technical approach to use. As a result of the ongoing conflict and lack of progress, two key team members left the company out of frustration. The project never got off the ground.

Neither of the above teams reached maturity. The team managers failed to take specific team-building actions such as *clarifying the team's mission* (situation A) and *establishing a consensus-making procedure* (situation B). These actions might have assisted the teams' development. It is important to recognize that any team member, not just the manager, could have suggested and helped to implement these actions.

---

[2] Parts of this chapter were adapted from David Wilemon, "Developing High-Performance Matrix Teams," in David I. Cleland (ed.), *Matrix Management Systems Handbook* (New York: Van Nostrand Reinhold, 1984). The quotes and examples used in this chapter are taken from field interviews. This chapter is the result of an ongoing research study by the authors which focuses on managing various types of teams in technology-based organizations.

Team building can be viewed as an ongoing responsibility shared by the team as a whole. Nevertheless, the team manager's authority and power usually puts him or her in a position to establish team building as a worthwhile enterprise. Team building requires a considerable amount of time, energy, commitment, and hard work; but the results can be enormous in terms of fulfilling performance, morale, and productivity goals.

Characteristics of a Mature Team

Successful team building helps the team develop into an optimally performing or "mature" group of colleagues. Some of the characteristics of a high performing mature team are[3]

- Team members manage their work against goals.
- Form follows function—the nature of the task at hand determines the team's procedures for doing the task.
- Decisions are made nearest the best source of information. All members may participate in and endorse the decision, but greatest weight is given to those team members who have the most expertise.
- Communication within the team is undistorted—people speak up without fear and are listened to without misperceptions; for example, there is no self-protective hedging or watering down of statements.
- Conflict is focused primarily on ideas—not people. Conflict is constructive—it helps reveal new information.
- Each person counts.
- The team learns through feedback among members and from the outside.

A FRAMEWORK FOR TEAM DEVELOPMENT

Table 9-1 contains our framework for team development. It describes four phases of task development and four phases of interpersonal relations development. In addition, typical examples of team-building actions used in each phase are briefly identified. Each of these team-building actions will be more fully explained in subsequent sections of this chapter.

The two sequences of phases are pictured in Table 9-1 as if transitions are exactly parallel; the "climate-setting/initiating" phase of task development corresponds to the "inclusion phase" of interpersonal relations, and so forth. Actually it is more realistic to think of them as approxi-

[3] Adapted from Richard Beckhard, *Organization Development: Strategies and Models* (Reading, Mass.: Addison-Wesley, 1969).

Table 9-1. A Framework for Team Development.

TEAM DEVELOPMENT TASKS	PERSONAL ISSUES FACING TEAM MEMBERS	EXAMPLES OF TEAM BUILDING ACTIONS FOR EACH DEVELOPMENT PHASE
	PHASE I	
Climate Setting/ Initiating What's our mission? How will we get started?	*Inclusion* How will I fit in? What will the other team members be like?	Clarifying the team's mission "Signing on" and including members Establishing a positive climate Building team identity
	PHASE II	
Goal Setting/ Work Planning What specific goals need to be developed? What roles need to be performed? What team procedures do we need?	*Power & Conflict* Will my contribution count? How can I influence the team? How can I maintain my identity & also be a "team player"?	Identifying goals Assigning & negotiating roles Identifying needed team procedures Determining how the team will relate to the "external world," e.g. management, the sponsor, functional groups Balancing team and individual recognition Managing conflict
	PHASE III	
Implementing How do we keep up momentum & team morale? How do we maintain performance standards? How do we stay on track?	*Personal Performance Within a "Larger System"* What standards will I try to meet: mine? the team's? the organization's? How do I keep up my energy & interest? How do I relate to the team manager?	Assessing the impact of team norms on performance Assessing on-going performance Encouraging lagging performers Keeping team progress visible Reaffirming team-leader trust
	PHASE IV	
Evaluating Team Results/ Following Up Did we meet our performance expectations? How do we sell team results? What are our next steps?	*Dealing With Success/Failure and Transition* How has my contribution been evaluated? What have I learned? How satisfied am I? How do I let go of the team?	Reviewing team progress Sharing experiences and feelings Consolidating the learning Recognizing individual contributors Celebrating Providing closure

mately parallel. It often happens that the speed of development is somewhat faster on one side of the framework than on the other. Although phase progression usually proceeds without any major reversing or phase-skipping, lingering issues can emerge. For example, someone who once felt included may later express feeling excluded, just as a team nearing completion of its task may decide go back and change a particular objective or procedure.

The framework is consistent with what managers and team members have said about their successful teams. We will use our framework to describe each phase of development, along with specific team-building ideas for managers to consider. The framework also applies, in microcosm, to team meetings. Suggestions for planning and running meetings based on the framework will be given in a later section.

Before the Team Begins

Team managers about to start new projects can help their teams most by learning everything they can about the task ahead and its objectives. Managers can also conduct research on organizational constraints and supports. For example, can the team count on the best equipment in the lab? How and when will progress reports be made to other teams? Once produced, will the results have to be "sold" to another part of the company? Are there organizational politics to be sensitive to?

Finding the answers to these and similar questions will help get teamwork started in the right direction. Knowing this information from the beginning helps team members feel confident that their manager can, if necessary, manage the organization to protect and benefit the team. The manager's actual ability to do that may depend on other factors entirely, but nevertheless having facts up front helps team members to feel secure. Team managers who seem to scramble for information or learn it by trial and error can present the opposite image, which can demoralize their teams.

Before the teamwork begins, managers can research the people who will be on the team. What are their backgrounds, competencies, and reputations? What advice for working with them is offered by former colleagues and managers? The goal is to obtain positive feedback and information. The surest way to invite trouble is to solicit negative data. Team managers can take an important early team-building action in their own minds by beginning with positive expectations of each person who will be on the team.

With positive, useful data about prospective team members, managers can begin to do some very effective team building in informal one-on-one meetings. They can praise individuals for competencies others attributed to them and express confidence in their abilities to contribute to the team. They can ask for input into preliminary planning and communicate willingness to listen. The ideas and information coming from such discussions are likely to be useful, but the sessions are particularly valuable for reducing team members' anxiety concerning such questions as: Why was I chosen? How hard will the work be? Can I trust you?

Managers can also use these sessions to begin their own personal relationships with team members. Just as a wise teacher finds something personal to remember about each child, an effective team manager finds something personal to share with each team member. That can simply mean rooting for the same football team or having children the same age. These become the conversation starters and comfort builders for the times ahead when one-on-one meetings will be held for other purposes.

We have pointed out the manager's need to establish early credibility so as to help team members feel comfortable and supported. These are important concerns for team members as they struggle to find a way to work together. These concerns, although often unvoiced, typically dominate the dynamics of the first phase of team development.

We will now carefully examine each of the four team development phases described in Table 9-1. For each phase we will describe the team's major tasks as well as the interpersonal concerns experienced by team members.

PHASE ONE: CLIMATE SETTING, INITIATING, AND INCLUDING

Phase one begins when the team officially meets to begin work. It ends when certain important norms have been set to create the team's climate, agreement has been reached on the overall mission, and all the team members have signaled they feel valued and included. In other words, they have signed on and committed themselves to the team and its task. These things take longer than most people believe, which means that team members often want to get started on the actual work very quickly without sufficient planning or cohesiveness. Nevertheless, it is important for team managers to resist the pressures to begin immediately on the work itself in order to do the team building for climate setting, mission clarifying, and inclusion that will underpin the eventual success of the team. Neglecting to do so can cause the team to fall apart, as the following case illustrates.

> A team formed to implement a management information system in a large manufacturing organization experienced trouble from the start. In

team meetings two members of the team dominated the discussions. Although the project leader had a printed agenda, it turned out to be useless since the meetings, once started, quickly got off track. The less assertive members eventually found excuses for not attending the team meetings—morale and team productivity dropped.

Climate Setting

It is easy to imagine how team members in the above case felt when they decided to drop out. The team's climate obviously was not one that encouraged members to express their feelings openly and directly. It is disappointing to witness the dissolution of a team before it even gets off the ground. It is particularly regrettable when you recognize that people who tend to be very talkative in meetings often do have the best ideas, and they do not necessarily have a great need to dominate others.[4] Typically such people have a poor sense of time passing and a habit of getting caught up in their own enthusiasm. They are often grateful to a team manager who helps them overcome those problems and they often respond well to simple guidelines for ensuring everyone's participation in meetings. Helping the team to create such guidelines is an example of a team-building action for positive climate setting.

Without formal team building the team climate usually emerges tacitly. That is, the norms that make up the climate are established with little real notice or discussion. For example, the climate will be more formal if the team manager sits at the head of the table instead of along one side or if team members wait to be recognized instead of speaking spontaneously. Status differences may be highlighted or ignored as a function of how people address each other. These are important climate-creating events—although they are rarely spoken about openly. Rather, people do whatever is consistent with the culture of the organization.

Climate setting is a team-building activity for purposefully creating a positive team climate based on norms the team selects for itself. Climate setting is also one of the most used and valued applications of team building in progressive organizations. One team manager in a major pharmaceutical company makes it a standard practice at the start of a new project to spend three or four days with his team dealing with climate-setting issues. His objective is to anticipate and avoid likely team problems and conflicts, and to establish approaches for handling such matters should they develop. The team undertakes several strategic planning exercises to minimize future detrimental surprises.

[4] Ivan D. Steiner, *Group Process and Productivity* (New York: Academic Press, 1972).

For those team managers with less generous budgets and too-close-for-comfort deadlines, less elaborate climate-setting processes are available. Another manager in a large electronics firm makes it a standard practice to begin every new team project with a two-hour discussion of "what if" questions. Both task and interpersonal concerns are raised. Some other questions that can be used for climate setting are

- What concerns do you have about this project?
- What concerns do you have about your role?
- What expectations do we have of one another?
- What do we need to meet these expectations?
- What issues do we need to discuss to get this project off to a successful start?

Questions such as these encourage people to express their hopes and fears at the start of a new team venture. The process establishes the positive norm that it is permissible to express one's ideas and feelings.

Follow-up questions are those which lead to actual planning. For example:

- How will we settle controversies?
- How will we handle someone's nonparticipation?
- How will we check for consensus on decisions?

Again, the opportunity for team members to relieve anxiety is just as valuable a benefit of climate-setting discussions as are the ideas and plans that result. It is also important to recognize that the team can reexamine and change its norms at any time. Periodic climate review sessions are recommended for ongoing team building. A managing engineer in a leading producer of reprographics equipment sets aside time in every meeting for the team to discuss questions such as, How satisfied are we with the way things are going? What changes would improve the way we work together? In short, team building for climate setting is most essential in the first phase of team development, but it can also be very helpful throughout the entire life of the team.

Task Initiation

Climate setting and initiating the task are not wholly separate, but *task initiation* specifically refers to defining the team's mission. A clear *mission statement* is essential for the team to fully grasp its overall assignment. We strongly suggest that team managers avoid simply announcing the mission and instead use a more consultative process. Announced rather than consultatively defined mission statements, however accurate, are often the targets of immediate attack. As one manager put it:

> In the beginning you bring them in, tell them what the challenge is, what the corporation would like. You have to work the mission out with the team. You can't tell them what has to be done, because, at that point, they're going to put up resistance and they're not going to be that committed.

We therefore suggest that managers charge the team with composing its own mission statement. Participation in that effort will do a great deal to promote each member's "ownership" of the task ahead.

Using a consultative process to define a team's mission is particularly useful when managers do not have "clear sailing orders." In this case, clarifying the team's assignment may require everyone's input.[5] This is particularly likely when the team is a committee of specialists brought together quickly to respond to a pressing problem. Beneath the presented problem, there is often a deeper issue that will require extensive teamwork and team building just to uncover.

Getting much further along in task initiation than the overall problem or mission is usually not advisable since the team has yet to address inclusion issues. Important decisions made before everyone feels a part of the team will probably only be "unmade," since "unmaking" decisions is an easy way for people who feel excluded to make their way into the team.

The ways by which team members first include themselves in whatever is happening sets very important precedents for them and for the other team members' expectations of them. For example, people may continue to use "decision unmaking" as their primary mode of relating to the team. One team manager recalled an individual who was very quiet throughout several meetings until consensus was about to be reached on a certain procedure, whereupon he finally spoke up, only to badger the team about a particular detail, creating a lengthy discussion until he could be appeased. For the rest of the team's existence he became known as the "nitpicker."

Including People

Team building for inclusion prevents the negative effects of disruptive entry into the team process, and it is probably the easiest kind of team building to do. It begins very simply with the introduction of team mem-

---

[5] Lowell W. Steele, "Managing Temporary Organizations," in *Innovations in Big Business* (New York: American Elsevier, 1975).

bers at the first meeting, which can be nicely handled by having everyone talk briefly about their backgrounds. After each person speaks, the team manager adds a few complimentary facts from his or her own research. This allows each team member to enter the team process in a very positive way and quickly reveals everyone's areas of expertise.

An important aside to mention here is that most people have trouble remembering names. After a certain amount of time they are embarrassed to admit that they do not remember someone's name. Especially in larger teams, meeting for the first time, it is helpful to have name plates (or at least name tags) available. Managers should make it a point to address people by name, to help themselves and everyone else begin to commit names to memory. People are often irritated when they are addressed by the wrong name after they feel everyone, especially the manager, should know who they are. Often team managers and members go to unusual lengths to avoid people whose names they have forgotten, with obviously detrimental effects on team functioning and morale, since the unknowing "unknowns" typically think they must have done something wrong. Having an identity within the team is important—it is the need that inclusion is meant to satisfy. Inclusion focuses on individuals. Its objective is to help each person feel counted. Another way to include people in the team is to ask each team member to complete sentences such as these:

- My favorite way of working on this kind of project is __________.
- An experience of mine that will be useful to this team is ________.
- The way I see myself fitting in is ________________________.
- During meetings I tend to ____________________________.

Any number of strategies can be used for inclusion, as long as they meet two criteria: (1) everyone participates about equally, and (2) people say something personal about themselves.

It takes a certain amount of skill to do this kind of team building without appearing condescending. Managers should avoid saying anything that would make them seem parental, such as elaborate thank yous or overgenerous compliments. The "team manager as parent" is a mode many get themselves into—it is very appealing to the more dependent members of the team but alienates the counterdependent and independent members.

Other team-building actions for inclusion throughout phase one and all subsequent phases are simple kinds of "gatekeeping" techniques that invite people back into the process whenever their participation has begun to lag. Some are straightforward, such as asking for someone's input on a matter during a meeting, or interrupting a temporary dominance by one or two talkative individuals for a round of brief comments by everyone. We believe that it is more effective to invite quiet people to talk than to ask talkative people to be quiet. Even the most diplomatic "gateclosing" request can hurt and demotivate people, often turning them into cynics.

It is a matter of judgment as to how much effort is appropriate to make toward shy people. A balance needs to be struck between protecting them and strengthening their abilities to contribute to teamwork. Sometimes it is advisable to ask people privately how they can be most comfortably included. Another method is to make shy people visible in other ways, for example by having them record things during team discussions or write something to be distributed with their name.

The most difficult people to include in teams are the "lone rangers" with very strong preferences for independent work. Far from being shy, they are typically very vocal in their disdain for meetings and teamwork in general; and they can immobilize a team. Their focus is on efficiency and perfection, qualities that are not likely to characterize the struggling, compromising, sometimes confusing style of group work. It would be to the lone ranger's advantage to learn to work with others, but if the transformation is too costly in terms of time and stress the following actions can help include them in the team:

- Let them work independently if necessary, and require progress reports and other kinds of team input.
- Keep in touch with them between meetings—drop by for visits and make phone calls.
- Reinforce the benefits of teamwork; for example, mention in their presence how a particular result would not have been possible without combined efforts. (Consciousness-raising)
- Include them in team rewards, such as official recognition from the organization and team bonus plans.
- Structure an assignment so that they have to depend on at least one other team member for a vital contribution.

The underlying strategy is to ease "lone rangers" into team participation by providing more controlled teamwork channels—for example, independent reports within the team—and rewarding cooperation and giving personal support for what are usually supersensitive egos.

When the "lone rangers" have finally been included in some significant manner, the team has become an entity unto itself—something more valuable than the sum of its parts. That event closes phase one, and is usually marked by an observable sign. Here are some examples:

- An enthusiastic chorus of agreement on some issue.
- Very low tension level—without exception, people obviously relaxed. More socializing and humor.
- Scanning eyes—at meetings people look around the group instead of only at the speaker or the person addressed.
- A noticeable physical rearrangement: chairs brought into a tighter circle—bodies leaning in.
- Suggestions for sharing team managerial duties.

To summarize, team building in phase one includes actions establishing the norms for interacting and procedures for handling forseeable problems (climate setting), clarifying the overall mission (initiating), and recognizing people's individuality and how to get them involved (inclusion). Throughout phase one, although there may be limited progress on the actual task, team members should feel a sense of accomplishment from organizing for the work ahead. The payoff is a good climate and a clear understanding of the mission. This creates a strong and durable foundation for phase two of the team development process.

PHASE TWO: GOAL SETTING, PLANNING, AND DISTRIBUTING POWER

The mission statement may need revision as the team moves into phase two because new aspects of the assignment or new, opportunity-producing capabilities of team members may have been discovered as a result of phase one team building. From this broad, comprehensive description of what the team expects to accomplish, specific goals need to be developed. These goals are for the overall assignment as well as for the subtasks. Task development in phase two begins with setting the goals and proceeds through the assignment of various responsibility areas (roles) among team members and designing specific task procedures. All this can go quite smoothly, or there can be interference of the dynamics if interpersonal relations development becomes stormy. The potential for that to happen gives this period of the team's life the often misunderstood label of *power phase*.

Once people feel that they belong, they look for ways to influence the team. That change can have a very positive impact on the team, actually propelling both task and interpersonal development forward. Depending on the personalities and histories of the team members, however, the change to wanting power rather than mere membership can create intense interpersonal conflict. The best of team managers may not be able to prevent this from happening, but they can do a great deal to counter the negativeness. Team-building ideas for doing that will be presented after those for task development, because how the task is managed has an important bearing on the potential for conflict.

Goal Setting

One team-building approach for developing goals is to list the proposed goals at a team meeting. This can be done with the total team or with subcommittees responsible for a particular task or a technological subsystem. For inexperienced teams or for goals that are particularly hard to clarify, it can be helpful to visualize a "solved state" or a desired feature of the result or product. Alternatively, teams can imagine a "worst-possible unsolved state" and then work backwards to focus on exactly what they want to happen to attain the goal under discussion.

When goals are very complex, they lose meaning; therefore, it is important to restructure them to reduce their complexity. This ensures a commonly shared perception about the goals, without which confused team members may end up working at cross-purposes. The following criteria can be used in evaluating the team's goals:

- Are they specific?
- Are the goals clear?
- Are they measurable?
- Are they time-based?
- Can they be communicated easily?
- Can they be clearly assigned to team members?

Agreement on these and any other criteria for goals, incorporating costs, should be reached before goal setting begins.

Goal agreements are instances of compatibility, whereas goal disagreements are instances of incompatibility. Goal disagreements do happen, especially on complex projects, and the team must work through the area of incompatibility to find a compromise or abandon one of the goals involved. If goal disagreement cannot be handled in a single meeting, an action plan needs to be developed. The action plan should identify who will resolve the goal disagreement, what process will be used, and when a resolution will be required.

The process of developing and reaching a consensus on goals is a cornerstone of effective team functioning. Unclear, or otherwise inadequate, goals lead to confusion, wasted motion, and stress. Team building for goal setting has a direct and important effect on interpersonal relations. This effect is even more prominent after the goals have been clearly identified and agreed on, and attention becomes focused on the roles and responsibilities of individual team members.

Dispersing Roles Among Team Members

Team building for this part of the planning begins with identifying and ends with dispersal of roles. in other words, these are the decisions about who does what in the continuing task development. Roles include both the formal and informal agreements and understandings of the responsibilities each team member has, and they should be clarified and delineated to prevent gaps or overlaps. One manager explained the importance of clear role definition this way:

> In R&D projects like mine you have a lot of different groups that must be coordinated to achieve project objectives. We have five different labs we're dealing with, as well as our client. In addition we have two subcontractors supporting us. I have to get clear on each group's responsibilities. Unless I do this, I'm going to run into some show-stopping problems real quick.

As indicated by this comment, the clear articulation of roles is necessary to avoid conflict or ambiguous situations which leave the team assignment vulnerable.

The following team-building questions help to identify, clarify, and negotiate roles and expectations:

- What roles should be performed to achieve our goals?
- Who should perform these tasks?
- What skills are needed? What skills do we have within the team?
- How will we deal with deficiencies?
- How much freedom should individual team members have in carving out their own roles?
- How will individual roles be coordinated to achieve a unified team effort?
- How will work assignments be allocated?
- What are the gray areas in defining responsibilities?

Answers to the preceding questions can also help to identify areas of agreement and conflict among team members about their roles, as well as areas where more information is needed. Once a conflict has been identified—for example, two team members want the same role or someone wants to avoid a role—new or changed roles can be negotiated. Role negotiation is a process whereby team members articulate their needs to those on whom they depend. One approach to role negotiation is to guide the entire team through the following steps:

- Each member develops a list of major expectations of other team members.
- These expectations are communicated to the relevant team members.
- Team members negotiate differences.
- New understandings are evaluated and tested for workability and sufficiency. Inadequacies are renegotiated.
- A summary of the conclusions is reached, and the processes used to produce them is proposed. The summary is then provided to the total team.

A simpler version of team building for role negotiation amounts to an exchange between two team members. In that exchange, the sender communicates specifically what is desired from the other person. These three statements can initiate the negotiation process between them:[6]

1. You help me be more effective if you do the following things.
2. You also help me be more effective if you do not do these things.
3. I hope you will continue to do the following things, because they help me be effective.

Role negotiation helps members create and adapt their expectations of one another. It is a democratic process, which demonstrates the team manager's trust. It remains the manager's responsibility, however, to make sure that no vital pieces of the team's task fall through the cracks of this process.

Developing Task Procedures

After roles are sorted out, planning team functioning will help to save time and achieve operations running smoothly. One manager discussed the usefulness of clear procedures as follows:

> Good procedures leave you time to plan ahead, manage the day-to-day nonroutine affairs, and focus on trouble spots. Procedures also eliminate a lot of conflict because they give the team a road map about how to deal with a lot of important issues.

[6] Roger Harrison, "Role Negotiation: A Tough Minded Approach to Team Development," in W. W. Burke and H. A. Hornstein (eds.) *The Social Psychology of Organization Development* (La Jolla, Calif.: University Associates, 1972).

The questions raised below illustrate the importance of having clearly defined procedures.

- How do we respond when some people have bigger work loads than others?
- How can we synchronize two subtasks in progress?
- How will interim progress reports be made?

Other procedures can be formulated in the technical, procurement, financial, personnel, and legal areas.

An excellent team-building practice is to hold as-needed problem-solving sessions to create new procedures or make changes in existing ones. Input from team members with relevant expertise can save the team's time in "wheel reinventing."

Decision Making

Implicit in the planning process is the need for team decision making. There are few things more unproductive than having an entire team try to make decisions that should be made at the individual level, or by a small group of specialists. When the technical quality or accuracy of the decision is at least as important as, if not more important than, its acceptability, then the decision should be made by only those individuals with appropriate authority and/or expertise. That may be the team manager alone or someone at a higher level or team members with the required credentials. Others inside or outside the team may be consulted or used as sources of additional information, at the discretion of the decision makers themselves.

At the opposite extreme of complexity, essentially trivial decisions should also be made at the individual or subcommittee level, or by any other expeditious means. Only substantive matters about which members are equally knowledgeable should be decided by sharing information and ideas within the entire team. The following six steps can significantly aid decision making, whether conducted by individuals or groups:

1. Assess the situation and collect or develop information that will be needed to make the decision.
2. Visualize, describe, or otherwise define a clear goal and/or set of criteria for the decision outcome.
3. Develop a variety of alternatives, suspending initial evaluation to avoid jumping to "quick fix" conclusions.

4. Weigh each alternative for feasibility, advantages, and disadvantages.
5. Select the best one in light of the criteria from step 2.
6. Test and evaluate the outcome and make adjustments (or choose another alternative) if warranted.

At points during the process, especially during the idea-gathering steps, brainstorming and "what if" sessions can be excellent vehicles for eliciting varied and creative responses. Other ways of creating a nonjudgmental atmosphere to assist the generation of ideas and uninhibited exchange of opinions are[7]

- Circulating brainstorm technique: Individuals list their ideas and read one per turn while a recorder assembles a master list.
- Nominal group technique: a variation of the above procedure but with the collection of lists without identifying sources.
- The Japanese "Ringi" technique: An extensive document of opinions and ideas is circulated for anonymous editing by team members until everyone is satisfied.
- The Delphi technique: a series of written questionnaires to which team members respond anonymously. Results are circulated until a consensus has been reached.

Recently, managers of technology-based organizations have become more concerned about team innovativeness and have encouraged more unusual methods of generating and examining ideas:

- The "odd-man-in" technique: inviting a "guest" from a very different field to expand creativity directly by introducing a new perspective or indirectly by jarring team members from their habitual ways of thinking.
- Fantasy, metaphors, analogies, and other ways of symbolizing the decision and examining it from an anxiety-free perspective.

Creating input for team decisions and reaching consensus can be aided by any of these methods or by combinations and variations of them that suit the needs and comfort levels of team managers and members.

Team-building actions for goal setting, role assignment and negotiation, procedure planning, and decision making are all part of task development

[7] Alvin Zander, *Making Groups Effective* (San Francisco: Jossey-Bass, 1983).

in phase two. None of these functions actually *must* cease at the end of phase two. Team members commonly tear apart and reset goals, trade and consolidate roles, abandon procedures, and "unmake" decisions right up until time runs out. Then they simply operate with the last set of circumstances. Many perfectly sound elements may be "thrown out with the bath water", but the never-ending currents of development in interpersonal relations create the waves of change, rarely with the team's awareness. These currents, strongest during this power phase, can build or destroy the team's cohesiveness.

As the team begins to define goals and subtasks, hidden agendas begin to surface: "I want to use this assignment as an avenue to promotion"; "This is my chance to get even with so-and-so for outmaneuvering me on that last project." Motives such as these remain hidden, sometimes even from their possessors, yet they underlie much of the competitiveness which characterizes phase two.

Power can mean more than just the capacity to influence the team or move its work in a certain self-serving direction. It can also mean gaining the respect of the team and having a special niche for making a positive contribution. This concept of power does not treat power as a total quantity to be divided into shares. Power is considered something essential for everyone to have as much of as possible, with no limit to the total *positive* power available to the team. This concept creates many effective team-building opportunities for team managers.

Distributing and Enhancing Power

Team building which applies the positive concept of power requires from the outset the team manager's awareness of his or her own role as model for the others and as a provider of rewards. To the extent that the team is treated as a unit and rewarded as a unit, members will feel empowered by their identification with the team. They will tend to minimize self-serving and competitive behavior, and see their own goals as subordinate to the team's goals. Throughout the power phase, then, it is important for the manager to address and praise the team as a whole, and to speak about the team to outsiders and to team members as if the team were an alive, achieving being. For example: "We're making great progress"; "This team is doing things that will make the rest of the company take notice"; "Thanks to all of you for coming so well prepared."

An even more subtle way to promote team identity is for the manager to use the scanning eye-contact pattern mentioned before as one of the physical signals of inclusion. The manager's visual attention is rewarding in and of itself, and continuing to look around at all the faces while speaking to the team distributes that reward evenly to everybody.

Other actions that promote a feeling of unity are well recognized—for example, creating some kind of badge of membership, either actually or symbolically. We know of one division of a computer graphics company that calls itself "the Black Sheep" because of a preference for using unorthodox procedures and occasional humorous rebellion against minor company policies. It is also one of the most productive and highly prized subunits in the company.

Social events involving the entire team can be important, especially when they include some words of recognition and other official cheerleading. One team member's opinion of the importance of such rewards was this:

> On a scale of 1 to 10, rewards are about 9.8. You get paid, which is fine, but you can also be told you are doing a good job—and that doesn't cost anything. For the latest project the whole team got together, a cross-section of groups which we didn't even know were working on the project, and at the dinner the plant manager recognized each group and what they had done for the plant. The verbal recognition in front of our peers was far more important than the dinner.

During the day-to-day managing of a team in phase two there are many other approaches to team building which aid team identity and discourage competitiveness, all of which combine rewarding the team as a whole with deliberate steps to "de-individualize" member contributions. For example:

- Disassociate ideas from their originators: "It's been suggested that we. . . ."
- Ask disagreeing team members to summarize each other's arguments.
- Collect members' ideas on cards and report them altogether and anonymously.

Where the balance should be struck between team and individual emphasis is a matter of judgment. Use mainly team rewards and treat the team only as a unit when conflict is destructive. Combine team and individual rewards and acknowledgment when conflict is neutral or constructive.

Flaring tempers, the formation of warring cliques, and extreme polarization interfering with teamwork are sure signs of destructive conflict;

minor bickering between team members, tolerated with amusement by the others, is more likely to be helpful than harmful. Conflict, after all, can be one of the team's chief ways of unearthing new ideas and information. Those involved strain to find arguments to support their views and often discover innovative solutions to team problems. Conflict is also many people's way of energizing themselves, as in the familiar expression "I'd rather have a strong enemy than a weak friend." Savvy team managers encourage conflict at times when it is needed:

> I'm not sure we really want to minimize the conflicts, because sometimes they inspire individuals to do things that normally they wouldn't accomplish. If I were to put a team together I may very well want two people that I know don't get along that well, but are good in their fields, just to keep things honest.

A common misperception about conflict is that it usually stems from personality differences—"laid back" people can't work with "aggressive" people, and "emotional" types can't work with "rational" types. More often it comes from system factors. Asking the following questions will help to uncover the real source of conflict between two team members:

- Are they held accountable for different things, for example, quantity versus quality?
- Is one rewarded for a certain behavior and the other for its opposite, for example, autonomy versus conformity?
- Are they expected to share scarce resources, for example, materials, support staff, money?
- Do they follow different informal norms; for example, do standards for performance differ in their two "home" departments?
- Have their backgrounds given them different perspectives?
- Do they understand each other's jargon?

Other sources of conflict were illustrated by a team manager in the following comment:

> A lot of time personality conflicts run deeper than just personality. Either it comes from an individual's lack of expertise in some areas which others recognize as a problem, or an individual feels he can do everything himself and wants to take control of the situation. You can modify it and make it better with a new viewpoint. It's good to have a cross-section of young minds with good tools to develop their new ideas

and older people with practical experience to define limitations and what you are really striving and working toward.

The mix of ages and career levels advocated above would probably produce a team bound to experience conflicts, but the manager quoted appears to have a very positive viewpoint toward conflict, and probably considerable skills in the management of it. Managers like this use a very positive concept of power and power differences in their teams.

Managing Conflict

It is extremely difficult to bury a destructive conflict. Suppressed or disguised conflict can actually do more damage since team members resort to the "strategies of the powerless."[8] For example:

- Team members will begin manufacturing excuses to avoid important team meetings.
- Team members may consciously or unconsciously sabotage team progress, by "misunderstanding" instructions and "forgetting" procedures.
- There will be loud grievances over unimportant issues and similar kinds of turf-protecting and bottlenecking activities.
- Overall progress will falter as interest wanes and people extend only halfhearted effort.

Settling a conflict to produce the best possible outcome for all concerned requires sensitive team building, some of which can be conducted during team meetings. Remember the team-building actions for distributing power, such as separating ideas from their originators and asking disputing team members to summarize each other's positions. Others include the following:[9]

- Explain how valuable differences of opinion can be in eliciting more options for the team to consider.
- Summarize both views without evaluation, to compensate for the failure of people in conflict to really listen to each other.
- Keep the meeting focused on important issues (conflicting team members will often avoid their uncomfortable feelings by dwelling on trivia and veering off onto tangents).

[8] Rosabeth Moss Kanter, *Job Power* (Cambridge, Mass.: Goodmeasure Inc., 1982).
[9] Rensis Likert and Jane G. Likert, *New Ways of Managing Conflict* (New York: McGraw-Hill, 1976).

- Control your own judgmental reactions to expressions of anger or anxiety, to make negative feelings more acceptable.
- Help the team develop procedures and ground rules for resolving conflict through negotiation.
- Provide open support and encouragement to all sides, to reduce tension and model fair treatment.
- Express positive feelings when reminding the team of past successes and expressing confidence in the future, so that temporarily unpleasant situations don't loom larger than they really are.

The conflicts hardest for most team managers to deal with are the occasional hostile verbal exchange. Restoring collaboration between two people who seem to be at war with each other is very stressful for most team managers, but the following steps can make the process easier.

In private meetings with each person:

1. Allow for the initial venting of emotion—listen and support without necessarily agreeing.
2. Ask open questions about incidents and facts contributing to the conflict (your actions included) and summarize the answers in order to communicate understanding.
3. Praise honesty, trying, and any other helpful developments.
4. Firmly emphasize the importance of team goals and the need for collaboration.
5. Ask for suggestions and praise those that seem feasible and fair.
6. Suggest taking the initiative to find a solution without your further interference.

If hostilities continue to block team progress:

7. Arrange a joint meeting in comfortable, neutral surroundings.
8. Reemphasize the need for collaboration and reduced hostility. Explain what may happen if there is no resolution, including the measures you will take.
9. Review the pertinent issues and get agreement on your summary's clarity and accuracy.
10. Insist on fair treatment and the discussion of tasks, responsibilities, and standards instead of personality characteristics and annoying mannerisms.
11. Mediate a dialogue aimed toward finding commonalities—summarize and clarify points as needed.

12. Ask for agreement on goals and a step-by-step action plan, in writing and signed, including a plan for monitoring progress.
13. Express confidence in the expected outcome.

Team managers, like most other people, prefer to work with others who enjoy and respect one another. This is not always possible, nor can it be forced. However, the above team-building process ensures the minimum amount of cooperation required to complete the task.

The essential theme of phase two's team-building actions is to meet individual needs for power by depicting power as accumulated *through* each other rather than *over* each other. Once the team has developed to the point where the work has been divided to everyone's satisfaction, both team and individual objectives are clear, and conflicts have been brought under control, the team is ready to move on to phase three. This transition is hardest to note, since it is not usually marked by an observable event. Typically there is a gradual lessening of tension and renewed enthusiasm. Keeping that vigor at a high level is one of the team manager's main challenges in phase three.

PHASE THREE: IMPLEMENTING, PERFORMING WITHIN THE SYSTEM

Task development in phase three involves doing the bulk of the work by following plans and procedures as established in previous phases. People settle into routines and complete the subtasks of their particular assignments, working at times with other team members and at times alone. Remaining unmet needs for power, that is, feelings of influence and value, may be handled by injecting new challenges into assignments and providing recognition. Perhaps more than most kinds of professionals, technology specialists are motivated by the challenge of the work itself, and keeping the challenges coming may be all that is needed to sustain high effort.

But the work can become tedious, and dealing with details is less enjoyable for many people than hashing out the big decisions. Additionally, some things will go wrong—customers will change specs, plans don't work, design snarls will resist untangling. Disappointments are inevitable and take their toll on morale. Finally, the last deadlines begin to look uncomfortably close, and expenditures will be nearing their prescribed limits, so that compromises have to be made. Not only are technical professionals motivated by challenge, a trait that may well carry them through all the setbacks, they typically have extremely high standards for their own individual work. Compromises involving technical quality are painful and increase stress. Sometimes the experience is severe, as in the following example:

> I've seen burnout in the sense that we exhausted all of our energy to accomplish something and we got nothing back. All we heard was that various things didn't satisfy the customer requirements. You have to recognize that some of the objectives, which are and should be set high, are not going to be met. People hate to hear that but there is a real world out there too. But the burnout came from putting in 110% effort and getting zip out of it.

Outside individuals or entities can be blamed, sometimes with justification, as in the common feuds between R&D and marketing divisions within companies. Top management, customers—a number of targets may be seen as sources of people's frustrations, including, of course, the team manager.

Depending on their proclivities for either compliance or rebelliousness, team members will vary in how they handle their feelings about authority issues. Some will act resigned, others will protest loudly, some will flourish on increased feelings of personal strength derived from the adversity and interdependency with other team members, including the manager. Experienced team managers have learned to take these differing orientations to themselves in stride. Two of them have made the following observations about their teams in phase three:

> The beginning is critical when you bring people together and motivate them to be a team. Then we are primarily concerned with the engineering phase. As the product moves into being built, team involvement fades away. Restoring team feeling by having frequent meetings is important. It's also important toward the middle to loosen control.

> To improve morale when things get tough and problems start to happen, the best thing is to have better communication, so that all the team members know what the next guy is doing and how well he is doing. Keep the progress out in the open, visible, and create milestones.

The above two quotations contain some very good advice for team building in phase three.

A very basic but tacit decision that will be made by the team in phase three is setting the final standard for the work. While the focus of conversation may be on time and money running out, underneath will evolve a new consensus on how well the work can and will be done—to what degree of excellence or adequacy. It is important for the team manager to ensure that standards are kept as high as possible, a responsibility likely to produce an intensified feeling of being under pressure. Much of that feeling may be justified as certain team members begin to show signs of actually slacking off.

Fear paralyzes—a phenomenon that disguises itself as procrastination. As some team members begin to worry about the final quality and acceptability of their contribution, as their frustration at not being able to do the high quality work of which they are capable increases, their behavior may look very much like psychological "bailing out" on the team. They may justify such withdrawal to themselves as political strategy—disassociating from a venture headed for failure, for example. Or they will suddenly find reasons to become more fervent about other responsibilities.

To avoid unpleasant confrontation of team "escapees" some team managers choose to become "martyred virtuosos"—for example, the "every night and all weekend marathon to get the report written" solo. A variation on the same theme is the covert refrain of "If you're so smart, you can do even more of it yourself." More team members will quickly, even gratefully, bail out. This is destructive to their own personal development and to the team's development.

A better policy for the team itself as well as for the manager is for him or her to join rather than take over the final phase of work if an all-out push becomes necessary. The need for even-handed and fair authoritative force may become necessary in some teams. For most teams, however, getting over the slump only requires an extra amount of encouragement and slightly more monitoring toward the end—checking in on progress every day, for example.

A final suggestion for team building during phase three is that team managers should attend carefully to their own needs for emotional support from people outside the team. It is extremely difficult to absorb the anxiety some team members are likely to vent without an extra measure of somebody's shoulder to cry on.

Phase three comes to a close when the task is complete. If the result or product begins to look good to team members as it assumes its final shape, expressions of feelings of relief and success will mark the start of phase four of team development. Teams that have failed, through their own shortcomings or for external reasons, at least will experience the relief. But neither the team's life nor its work is completely finished. Typically the team must now "sell" its result to the rest of the organization, its sponsor, or its customer. The evaluative reactions to the result, both inside and outside the team, set the stage for phase four.

PHASE FOUR: EVALUATING, FOLLOWING UP, LETTING GO

This chapter began with a list of characteristics of a mature team, the last item of which was "The team learns through feedback among members and from the outside." So many teams allow themselves to fall apart without more than cursory self-evaluation, cheating themselves of an important aspect of team maturity. As the final team output is assembled and made ready to carry outward for the judgment of others, an extremely productive team-building action is to set aside time for a thorough review of the team's history, both in terms of its task progress and its interpersonal functioning. This action will make possible the consolidation of everyone's learning.

Just as important as the work of reviewing and learning from its own history, the team must also progress through the final phase of interpersonal relations development. It is called the intimacy phase because members examine their feelings for each other and adjust their emotional ties to the team, drawing closer together or distancing themselves. More talking about personal matters—or a sudden dropoff of such conversations—reveals that this is happening. If this phase is blocked by early dissolution of the team, members may experience the sadness that usually accompanys unfinished personal development, even if the reasons are not consciously understood. Therefore, a review of the team activities can meet important psychological needs as well as reveal important learning.

This review does not require a great deal of structure or formality; in fact, such an approach could have a stifling effect. The idea is to have an open, friendly discussion about how things went and how team members feel. A few questions to get things started are:

- How well did we accomplish our objectives? What helped and what hindered?
- What were this team's main strengths and weaknesses in terms of (1) task performance and (2) the process by which we worked together?
- What were the events that impeded our team from functioning at a high, sustained level?
- How did our relationships with external groups affect the way we worked together?
- How did we help each other become more effective?
- How did the team's functioning affect the performance of each individual member?
- How did we increase our capabilities to diagnose and solve problems as our team progressed through its life cycle?

Most teams achieve at least some success. For these teams, the team evaluation becomes an opportunity to celebrate. Team managers should be wary, however, that shared success can have the effect on team members of not wanting to give up the team. This sometimes leads to undoing part of the task in order to do it over. Strange as it may seem, this happens rather frequently, as one manager remarked in the following comment:

> Nobody wants to quit. You get team members that are all motivated and there's three more things they'd like to do before they're done. So trying to get a project stopped is a difficult stage in the life cycle of a successful project.

This observation reinforces the need for very clear goals and criteria for "doneness."

Another common experience is to have team output ignored. This can be worse than having it rejected because rejection often includes criticism, which is at least informative. In many cases, indifference to a project can be traded to a team's failure to "sell" the result. Team outputs often end up being "pushed" by individuals, who simply get too busy or do not have enough influence to accomplish what needs to be done. Wise team managers realize that the entire team can have a continuing role in follow-up activities, including persuading top management or customers to accept the team's product or act on team recommendations.

Sharing credit is important but it can be difficult to achieve. Many well-intentioned team managers find themselves overassociated with the work of their teams. Our cultural norm is to recognize individual over team achievements; therefore, reports representing weeks or even years of combined contributions, for example, become labeled as SO-AND-SO'S report. Even the coverage of internal media, such as company newsletters, is commonly slanted in one person's direction. As the likely candidate to benefit from such publicity, a team manager who is a sincere team builder rather than an exploiter will take steps to ensure that credit is shared. That usually requires documenting or at least being very vociferous about individual contributions of team members.

Ongoing teams simply get on with the next project, but some kind of team rewarding event should be staged to demarcate the end of one life cycle and the beginning of another. Temporary teams face the more difficult task of final separation, although team members may stay in contact with each other. Closing ceremonies are important team-building events, because they satisfy team members' psychological needs for unambiguous endings and transitions. Ceremonies can be rather simple affairs. Even a last round of handshaking at the last meeting can be sufficient.

Whatever the scope or nature of the ceremony, it is the team manager's last team-building action—that taken for the purpose of expressing thanks and praising the team for its efforts, outcomes, and support. If team-building actions were incorporated into the team management process all along, then the team leader will be addressing a cohesive, highly productive group of specialists who have forged a high level of trust. "Thanks" becomes a very natural and familiar word for them to hear a final time.

Table 9-2. Outline for a Team Meeting.

PHASE	CHIEF CONCERNS OF TEAM MEMBERS	AGENDA ITEMS	REASONS
I	Fitting in Being accepted	Welcome Announcements Plan for meeting Praise Easy decisions	People need to feel that they belong Create early success experience Set positive meeting climate
II	Influencing Contributing Finding a role	Hardest decisions	People most motivated to contribute People most positive Disagreements often assist better decisions
III	Meeting personal, team, and external standards Relationship with the leader—e.g., dependence, resistance, autonomy	Easy decisions	Need to refocus on the team rather than on the leader or other individuals Need to reduce tension Need for "losers" of hard decisions to feel like "winners" again
IV	Closeness Personal matters Ties with others	For-discussion-only items Noting successes Praise for team Thanks	Need to promote team cohesion Need to encourage Need to end positively

MANAGING TEAM MEETINGS

Carefully managing its meetings is one of the team's most important opportunities to continue the team-building process and advance the progress toward team goals. We find many similarities between the team development framework presented earlier and the processes that a team can follow in managing an effective team meeting. Our suggestions for a team meeting framework are presented in Table 9-2. It contains four-phased meeting outline that also takes into consideration the main concerns people are likely to have during each phase, and the need to keep the team climate positive. Additional suggestions for managing team meetings are

- Model the behaviors you want others to display—namely, those that exhibit trust, respect, honesty, concern for quality, and proper responsibility to the organization and its customers.
- Address, speak about, and praise the team as a whole, especially when dealing with conflictful issues.
- Look around the group rather than at only one or two individuals as you speak.
- Keep the discussion on track, especially when tension is high.
- Separate ideas from people or connect ideas to people to discourage or encourage competition, respectively.
- Use humor to reduce tension but not to avoid tension-causing issues.
- Always begin and end a meeting with something positive.

SUMMARY

The increasing sophistication of our technology requires the use of teams to solve problems, develop new products, implement new manufacturing methods, and develop approaches for increasing technological productivity. As a consequence, technical managers will increasingly be required to manage teams of specialists—often drawn from diverse areas within the organization. The technical manager's challenge will be to help such a team of specialists develop into a high-performing work unit.

We believe our team development framework provides a number of useful, easily implementable processes to develop teams. Its strength is that it is sensitive to both the task requirements the team faces and the interpersonal concerns of team members. The results of successful team development can be high performance, team member satisfaction, and the growth of individual team members. We think these are all worthy and challenging goals for technical leaders.

BIBLIOGRAPHY

Blake, R, and J. Mouton, "Overcoming Group Warfare." *Harvard Business Review*, November–December 1984, pp. 98–108.

Cooper, C. L. (ed.). *Theories of Group Processes*. New York: Wiley, 1975.

Dyer, W. G. *Team Building: Issues and Alternatives*. Reading, Mass.: Addision-Wesley, 1977.

Francis, D., and D. Young. *Improving Work Groups: A Practical Manual for Team Building*. San Diego, Calif.: University Associates, 1979.

Golembiewski, R. T. *The Small Group*. Chicago: University of Chicago Press, 1962.

Gordon, W, and R. Howe, *Team Dynamics in Developing Organizations*. Dubuque, Iowa: Kendall/Hunt, 1977.

Katz, R. "High Performance Research Teams." *The Wharton Magazine*, **6** (Spring 1982):29–34.

Luft, J. *Group Processes: An Introduction to Group Dynamics*, 2nd ed. Palo Alto, Calif.: National Press Books, 1970.

Mahoney, F. X. *Team Development: From Definition to Operation*. New York: AMACOM, 1984.

Merry, V., and M. Allerhand. *Developing Teams and Organizations*. Reading, Mass.: Addison-Wesley, 1977.

Napier, R. W., and M. K. Gershenfeld. *Groups: Theory and Experience*. Boston: Houghton Mifflin, 1981.

Tuckman, B. W., and M. A. C. Jensen, "Stages of Small Group Development Revisited." *Group and Organizational Studies*, **2** (1977)419–27.

INDEX

INDEX

Behavior
  Argyris, C., Personality and organization, 125
  basic forces, 116–117
  Blake-Mouton managerial grid, 122–128
  Hawthorne Studies, 116–117
  Herzberg, F., Hygiene theory, 121–122
  Likert, R., Principle of supportive relationships, 124
  Maslow's Hierarchy of needs, 118–121
  McGregor, D., Theory X and Theory Y, 123–124
  organizational, 116
  Ouchi, W. G., Theory Z, 125–126

Communications, 37–76
  complexities of, 38
  cultural side, 64–66
  delivering speech, 72–73
  effective listening, 51–53
  interviewing, 73
  message misinterpretation, 48
  model of, 43–48
  nature of, 37–40
  networks, 41–43
  overcoming barriers, 48–51
  organizational patterns, 53–61
  philosophies, 74–75
  with customers, 63–64
  with executives, 61–62
  with subordinates, 62–63
  writing reports and memos, 70–72
Conflicts, 196–226
  environment, 197–200
  handling modes, 204–205, 214–215, 317–319
  intensity, 203–204
  managing, 201–203
  project life cycle stage, 204–208
  resolution of, 219–221
  understanding superior, subordinate, and functional, 221–225

Decision making, 227–263, 312–314
  certainty/uncertainty, 231
  choice elements, 240–244
  conflict in, 240
  controllable/noncontrollable factors, 231–233
  framework of, 230–237
  judgmental decisions, 230
  problems and opportunities, 229–230
  programmed/nonprogrammed decisions, 233–237
  rational decisions, 230–231
  steps in, 237–239
  within management functions, 244–253
Decision-making authority, delegation of, 239–240

Engineering teams, types, 2 (table)

Groups
  dependency, 34–35
  emotional support of, 33
  status, 33–34

Leadership, 77–111
  behavior pattern, 84–86
  cultural elements, 89–91
  definition of, 78–81
  factors in R & D, 104–106
  historical view, 77–78
  in 1980s, 93–94
  in project environment, 106–109
  life-cycle, 100–103
  participative, 88–89
  selecting appropriate style, 96–100
  studies, 94–96
  styles, 86–88
  traits, 83
  types and functions, 81–82
Linear responsibility chart, 184–191
  alternative applications, 191–194
  developing, 189

Management
  knowledge, skills, and attitudes, 22–24
  role playing, 25
Management science, 253–258
  approaches, 256–257
  data gathering, 255
  implementation, 256
  model building, 255
  orientation, 254
  problem definition, 254
  solution, 255
Managing creativity, 133–136
  applying programs, 135–136
  defining creative individual, 133–134
  motivating creative individual, 134–135
Meetings, conduct of, 66–70
Morale, 153–159
  building of, 158–159
  employee behavior, 154–156
  measuring of, 156–157
  nature of, 153–154
Motivation, 112–161
  applying theories, 127–131
  motivating people, 112
  nature of, 113–116
  problem employee, 136–139
  reward system, 139–142
  working with people, 131–133
Motivational process, 142–153
  employee rights, 150–151
  freedom, 148
  individualism, 149
  job enlargement, 152–153
  need for order, 151–152
  performance, productivity, and job satisfaction, 145–148
  work performance, 142–146

Organizational design, 172–182
  accountability, 182
  authority, 176–178
  delegation of, 174–176
  power, 178–181
  responsibility, 181
Organizational roles, 182–184, 167–168
Organizational teams, 26–27, 166–167, 170–171

Planning and organizing, 162–195
  definition of, 162–163

Professionals, 21–22
  transition to management, 21
Project engineering, 4–6
Project life cycle, 204–219
Project management, 3–4
  stress in, 274–275
  stress model, 277–281

Quality circles, 14–15

Stress and burnout
  categorizing feelings, 287–294
  emotions, 284–287
  introduction to, 272
  questionnaire, 275–277
  recognizing changes, 294–296
  relaxers/coping methods, 294
  understanding of, 273–274

Task force management, 15–17
Team
  building of, 189–191
  crisis management, 17–18
  goals, 164
  mature characteristics, 299
  managing meetings, 325
  member roles, 167–168
  objectives, 163–164
  organization structure, 166–167, 170–171
  planning, 169–170
  product-design, 14
  production, 6–11
  resources, 169
  strategies, 164–168
  style, 168
  "systems" support, 168
  worker-management, 12–14
Team development, 297–326
  before team begins, 301–302
  climate setting, initiating and including, 302–308
  decision making, 312–314
  distributing and enhancing power, 314–317
  evaluating, follow-up, letting go, 322–324
  focus, 298–299
  framework, 299–301
  goal setting, planning and distributing power, 308–312

implementing, performing within the system, 319–321
Team management, 1–35, 245–253
ambience of, 1–35
functions of, 245–253
Team membership, 27–33
acceptance, 29–30
behavioral norms, 31–32
benefits, 27–28
identification, 28–29
protection and assistance, 32–33
Time management, 264–296
effective, 268–270
forms, 270–272
time robbers, 265–268, 281–283
understanding, 264
Trade-off analysis, 258–260
methodology for, 260–262